METHODOLOGIES FOR ASSESSING PIPE FAILURE RATES IN ADVANCED WATER COOLED REACTORS

The following States are Members of the International Atomic Energy Agency:

AFGHANISTAN
ALBANIA
ALGERIA
ANGOLA
ANTIGUA AND BARBUDA
ARGENTINA
ARMENIA
AUSTRALIA
AUSTRIA
AZERBAIJAN
BAHAMAS
BAHRAIN
BANGLADESH
BARBADOS
BELARUS
BELGIUM
BELIZE
BENIN
BOLIVIA, PLURINATIONAL STATE OF
BOSNIA AND HERZEGOVINA
BOTSWANA
BRAZIL
BRUNEI DARUSSALAM
BULGARIA
BURKINA FASO
BURUNDI
CAMBODIA
CAMEROON
CANADA
CENTRAL AFRICAN REPUBLIC
CHAD
CHILE
CHINA
COLOMBIA
COMOROS
CONGO
COSTA RICA
CÔTE D'IVOIRE
CROATIA
CUBA
CYPRUS
CZECH REPUBLIC
DEMOCRATIC REPUBLIC OF THE CONGO
DENMARK
DJIBOUTI
DOMINICA
DOMINICAN REPUBLIC
ECUADOR
EGYPT
EL SALVADOR
ERITREA
ESTONIA
ESWATINI
ETHIOPIA
FIJI
FINLAND
FRANCE
GABON
GEORGIA
GERMANY
GHANA
GREECE
GRENADA
GUATEMALA
GUYANA
HAITI
HOLY SEE
HONDURAS
HUNGARY
ICELAND
INDIA
INDONESIA
IRAN, ISLAMIC REPUBLIC OF
IRAQ
IRELAND
ISRAEL
ITALY
JAMAICA
JAPAN
JORDAN
KAZAKHSTAN
KENYA
KOREA, REPUBLIC OF
KUWAIT
KYRGYZSTAN
LAO PEOPLE'S DEMOCRATIC REPUBLIC
LATVIA
LEBANON
LESOTHO
LIBERIA
LIBYA
LIECHTENSTEIN
LITHUANIA
LUXEMBOURG
MADAGASCAR
MALAWI
MALAYSIA
MALI
MALTA
MARSHALL ISLANDS
MAURITANIA
MAURITIUS
MEXICO
MONACO
MONGOLIA
MONTENEGRO
MOROCCO
MOZAMBIQUE
MYANMAR
NAMIBIA
NEPAL
NETHERLANDS
NEW ZEALAND
NICARAGUA
NIGER
NIGERIA
NORTH MACEDONIA
NORWAY
OMAN
PAKISTAN
PALAU
PANAMA
PAPUA NEW GUINEA
PARAGUAY
PERU
PHILIPPINES
POLAND
PORTUGAL
QATAR
REPUBLIC OF MOLDOVA
ROMANIA
RUSSIAN FEDERATION
RWANDA
SAINT KITTS AND NEVIS
SAINT LUCIA
SAINT VINCENT AND THE GRENADINES
SAMOA
SAN MARINO
SAUDI ARABIA
SENEGAL
SERBIA
SEYCHELLES
SIERRA LEONE
SINGAPORE
SLOVAKIA
SLOVENIA
SOUTH AFRICA
SPAIN
SRI LANKA
SUDAN
SWEDEN
SWITZERLAND
SYRIAN ARAB REPUBLIC
TAJIKISTAN
THAILAND
TOGO
TONGA
TRINIDAD AND TOBAGO
TUNISIA
TÜRKİYE
TURKMENISTAN
UGANDA
UKRAINE
UNITED ARAB EMIRATES
UNITED KINGDOM OF GREAT BRITAIN AND NORTHERN IRELAND
UNITED REPUBLIC OF TANZANIA
UNITED STATES OF AMERICA
URUGUAY
UZBEKISTAN
VANUATU
VENEZUELA, BOLIVARIAN REPUBLIC OF
VIET NAM
YEMEN
ZAMBIA
ZIMBABWE

The Agency's Statute was approved on 23 October 1956 by the Conference on the Statute of the IAEA held at United Nations Headquarters, New York; it entered into force on 29 July 1957. The Headquarters of the Agency are situated in Vienna. Its principal objective is "to accelerate and enlarge the contribution of atomic energy to peace, health and prosperity throughout the world".

IAEA NUCLEAR ENERGY SERIES No. NR-T-2.16

METHODOLOGIES FOR ASSESSING PIPE FAILURE RATES IN ADVANCED WATER COOLED REACTORS

INTERNATIONAL ATOMIC ENERGY AGENCY
VIENNA, 2023

Printed by the IAEA in Austria
June 2023
STI/PUB/2043

IAEA Library Cataloguing in Publication Data

Names: International Atomic Energy Agency.
Title: Methodologies for assessing pipe failure rates in advanced water cooled reactors / International Atomic Energy Agency.
Description: Vienna : International Atomic Energy Agency, 2023. | Series: IAEA nuclear energy series, ISSN 1995-7807 ; no. NR-T-2.16 | Includes bibliographical references.
Identifiers: IAEAL 22-01564 | ISBN 978–92–0–150322–0 (paperback : alk. paper) | ISBN 978–92–0–150422–7 (pdf) | ISBN 978–92–0–150522–4 (epub)
Subjects: LCSH: Water cooled reactors. | Pipeline failures. | Nuclear reactors.
Classification: UDC 621.039.58 | STI/PUB/2043

FOREWORD

The IAEA's statutory role is to "seek to accelerate and enlarge the contribution of atomic energy to peace, health and prosperity throughout the world". Among other functions, the IAEA is authorized to "foster the exchange of scientific and technical information on peaceful uses of atomic energy". One way this is achieved is through a range of technical publications including the IAEA Nuclear Energy Series.

The IAEA Nuclear Energy Series comprises publications designed to further the use of nuclear technologies in support of sustainable development, to advance nuclear science and technology, catalyse innovation and build capacity to support the existing and expanded use of nuclear power and nuclear science applications. The publications include information covering all policy, technological and management aspects of the definition and implementation of activities involving the peaceful use of nuclear technology. While the guidance provided in IAEA Nuclear Energy Series publications does not constitute Member States' consensus, it has undergone internal peer review and been made available to Member States for comment prior to publication.

The IAEA safety standards establish fundamental principles, requirements and recommendations to ensure nuclear safety and serve as a global reference for protecting people and the environment from harmful effects of ionizing radiation.

When IAEA Nuclear Energy Series publications address safety, it is ensured that the IAEA safety standards are referred to as the current boundary conditions for the application of nuclear technology.

The successful deployment of advanced water cooled reactor technologies includes the development of design certification probabilistic safety assessment studies and reliability and integrity management programmes for ageing management and in-service inspection. The consideration of piping reliability is an essential element of these probabilistic safety assessment studies and reliability and integrity management programmes. A well-recognized technical challenge in the development of probabilistic pipe failure metrics applicable to advanced water cooled reactors is the scarcity or lack of relevant operating experience data on which to base or inform the piping reliability parameter assessment.

This publication provides a comprehensive review of good practices for the assessment of piping reliability parameters for advanced water cooled reactors. Good practices are those processes and analytical tasks that would be expected in piping reliability analysis for the results to be realistic representations of piping structural integrity. Piping reliability is a complex subject that has been studied extensively and from various technical perspectives (e.g. from development of design rules to development of material degradation mitigation practices). To assist Member States in applying adequate methodologies to pipe failure rates analysis in advanced water cooled reactors, the IAEA organized a three year coordinated research project entitled Methodology for Assessing Pipe Failure Rates in Advanced Water Cooled Reactors (2018–2021). This publication builds on technical insights that have been obtained using different state of the art methodologies when applied in multiple analytical contexts and responding to the requirements of different national codes and standards.

The IAEA is grateful to those Member States that provided valuable support in the form of experts and technical information. In addition, the IAEA wishes to thank all the experts who participated in the drafting and review of this publication. The IAEA officer responsible for this publication was T. Jevremovic of the Division of Nuclear Power.

CONTENTS

1. INTRODUCTION

1.1. BACKGROUND

Piping integrity has been an important consideration throughout the development of the nuclear power plant (NPP) safety and reliability technologies. In 1963, the United States Atomic Energy Commission undertook a survey of piping failures for input to a primary coolant pipe rupture study [1]. Among the conclusions of the survey: "Piping systems which have been designed and constructed within established code criteria will exhibit high reliability, and a catastrophic failure (complete severance rupture) appears unlikely; however, less severe failures will obviously occur." At the 1965 International Symposium on Fission Product Release and Transport under Accident Conditions [2], reference was made to "incipient failures" in the form of "significant cracks" having been discovered in the piping of the main primary coolant system in two reactors in the United States of America. These early observations have been under detailed scrutiny ever since in order to advance piping reliability analysis methodologies [3–23].

Building on today's state of knowledge, this publication addresses good practices for piping reliability analysis to develop advanced WCR pipe failure rates. Piping reliability and its quantitative assessment is one of several important aspects of probabilistic safety assessment (PSA) and enters into multiple areas of PSA implementation and application, including assessment of loss of coolant accident (LOCA) initiating event frequencies, moderate and high energy line break frequencies, and internal flooding initiating event frequencies. Practical applications of piping reliability models include risk-informed safety classification for use in risk-informed repair/replacement activities, fitness for service assessments, risk-informed in-service inspection (ISI) programme development, and reliability and integrity management (RIM) programmes to optimize the selection of ISI locations and the implementation of non-destructive examination of piping. However, a well-recognized technical challenge is the scarcity, or lack of relevant advanced WCR specific operating experience (OPEX) data on which to base or inform piping reliability parameter assessment.

The successful deployment of advanced WCR technologies includes the development of design certification PSA studies. These PSA studies must address piping reliability in multiple contexts. Lacking OPEX data for the advanced WCR piping systems — with their unique designs, materials and operating environments — there is not yet a consensus on the technical approach for how to develop new advanced WCR centric pipe failure rates. The term 'centric' means that a derived pipe failure rate as much as possible reflects a specific piping system's design characteristics and operating environment.

To assist Member States in applying adequate methodologies to pipe failure rates analysis in advanced WCRs, the IAEA organized a three year coordinated research project entitled Methodology for Assessing Pipe Failure Rates in Advanced Water Cooled Reactors (CRPI31030), which took place from 2018 to 2021 with the participation of ten institutions from eight Member States. The TECDOC-1988 on Technical Insights from Benchmarking Different Methods for Predicting Pipe Failure Rates in Water Cooled Reactors as a final report of this coordinated research project was published in 2021. It summarizes the results of relevant benchmark examples and provides a technical basis for establishing NPP piping reliability parameters [24]. This publication then builds on these technical insights developed using different state of the art methodologies when applied in multiple analytical contexts and in responding to the requirements of different national codes and standards.

International organizations like the European Utility Requirements Organization and the American Society of Mechanical Engineers (ASME) are engaged in activities directed towards the establishment of technical requirements for the performance of PSA and structural integrity assessment for advanced WCRs. With respect to ISI of piping systems, ASME has rewritten parts of ASME Section XI (Rules for In-Service Inspection of Nuclear Power Plant Components) to account for lessons learned from ISI and risk-informed ISI as implemented for operating WCRs. The current edition of the ASME Boiler and Pressure Vessel Code Section XI was issued in 2019. Division 2 (Requirements for RIM Programmes for Nuclear Power Plants, BVPC-XI-2-2019) of ASME XI documents the requirements for the creation of an RIM programme for advanced nuclear reactor designs [25]. The RIM process addresses the entire life cycle for all types of advanced WCRs as well as non-light water advanced NPPs. The RIM process requires a combination of monitoring, examination, tests, operation and maintenance requirements that ensures each structure, system and component (SSC) meets plant level risk and reliability goals that are selected for the RIM programme. The RIM process consists of the following steps:

— RIM programme scope definition;
— Degradation mechanism assessment;
— Plant and SSC reliability target allocations originating from a plant specific advanced WCR PSA;
— Identification and evaluation of RIM strategies;
— Evaluation of uncertainties;
— RIM programme implementation;
— Performance monitoring and RIM programme updates to account for new OPEX and piping design modifications.

According to the new Division 2 of ASME Section XI, a reliability target is defined as "a performance goal established for the probability that an SSC will complete its specified function in order to achieve plant-level risk and reliability goals." With respect to PSA, ASME and the American Nuclear Society are developing a new standard for the conduct of plant specific studies acknowledging the technical/analytical challenges associated with the lack of OPEX data for advanced WCRs. The new standard is referred to as the advanced light water reactor PSA standard [26].

Considerations of piping reliability enters into multiple aspects of advanced WCR safety and reliability. Pipe failure rates (or probabilistic failure metrics) are input to, for example, PSA models, RIM programme development, fitness for service assessments and probabilistic analysis of pipe failure events to assess their risk significance. This publication addresses piping reliability analysis broadly. The scope of this publication is not limited to considerations of primary pressure boundary piping components (safety class 1 piping). Considerations of piping reliability for systems located outside the primary containment are also included. The advanced WCR piping material selections and piping system design philosophies imply higher or significantly higher levels of pressure boundary structural reliability. It is envisaged that probabilistic pipe failure metrics rates are at least an order of magnitude lower for advanced WCRs, as is indicated in Fig. 1, which has been adapted from [27]. Because of a lack of OPEX data on which to base a piping reliability analysis the resulting uncertainties would be expected to be significant. A technical challenge is to ensure that new piping reliability analyses are well documented and with sufficiently validated model inputs and outputs.

1.2. OBJECTIVE

The objective of this publication is to provide engineering reference information on applying adequate methodologies to pipe failure rates analysis in advanced WCRs. Addressed in this publication, the high level objective from the coordinated research project (CRPI31030) on Methodology for Assessing Pipe Failure Rates in Advanced Water Cooled Reactors is to provide Member States with open access to a strong basis for establishing advanced WCR piping reliability parameters consistent with required standards and relevant deployable advanced WCRs. The guidance provided in this publication, describing good practices, represents expert opinion but does not constitute recommendations made based on a consensus of Member States.

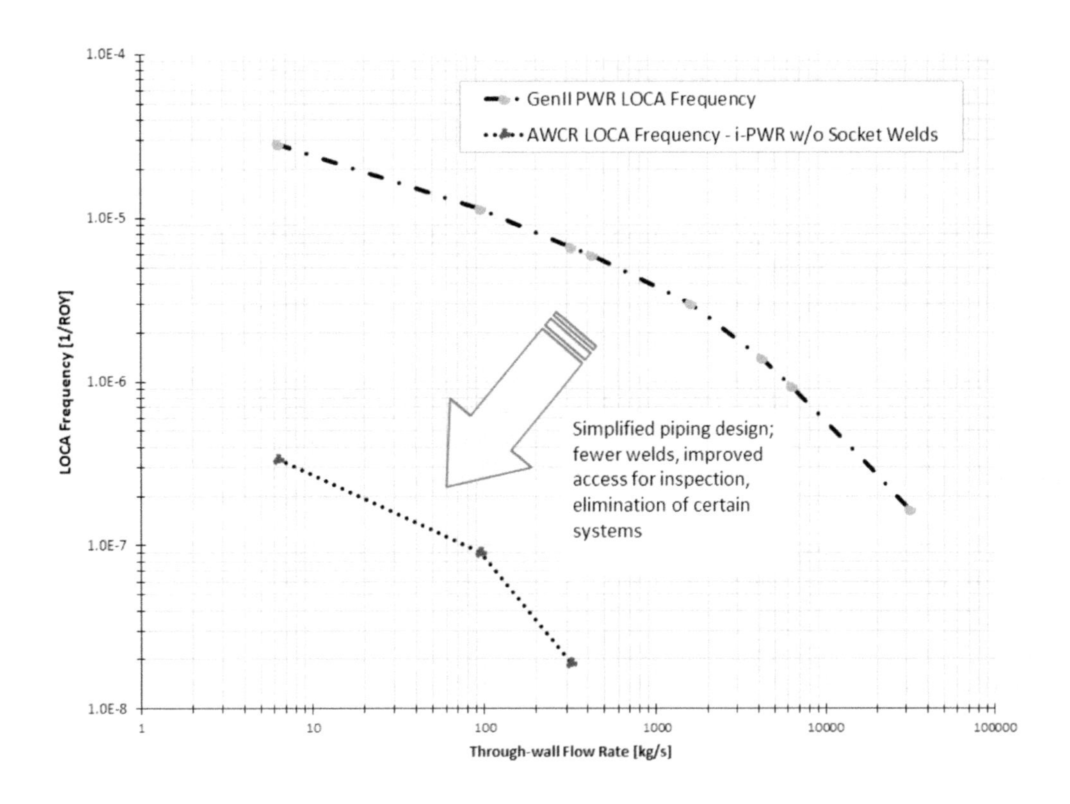

FIG. 1. LOCA frequencies in WCRs versus advanced WCRs [27].

1.3. SCOPE

This publication provides:

- Background on piping reliability analysis in different contexts, for example, PSA, design certification PSA, operability or fitness for service assessment, ISI programme development and optimization as part of RIM.
- State of the art information regarding piping reliability methodologies and their implementation.
- Technical insights into piping reliability analysis tools and techniques for advanced WCR applications.
- Documentation on the advancement of the existing piping reliability analysis methods across the Member States to explicitly address the factor of influence of, for example, new metals, non-destructive examination technologies and ageing effects on assessed pipe failure rates.
- Approaches on how to modify an existing set of piping reliability parameters originally developed for operating WCRs to be made applicable to advanced WCRs.
- Methodology for ageing factor assessment that utilizes the existing OPEX data. The methodology should account for the uncertainty in assessed increase or reduction in ageing factor as a function of plant age. The ageing factor assessment may include considerations of existing OPEX data, laboratory data, expert judgement and results from applications of structural reliability models.

1.4. STRUCTURE

This publication provides a guide to models and analysis methods and techniques for estimating advanced WCR pipe failure rates. Section 2 includes a high level summary of the piping OPEX from the late 1960s to the early 2020s and addresses WCR as well as advanced WCR piping material degradation mechanisms. In Section 3, a seven-step piping reliability analysis framework is introduced. Embedded in this framework are good practices that have evolved from practical analyses that have been informed by OPEX data, fracture mechanics and results from the material sciences research and development (R&D). The details of this framework are presented in Sections 4 through to Section 10. Suggested guidance on how to develop advanced WCR pipe failure rates are described in Section 11. The many different advanced WCR designs share a common piping system design philosophy in that they utilize materials that have a demonstrated high resistance to environmental degradation. The piping system layouts, fabrication and installation practices are evolutionary in that they favour structural robustness, simplicity and good accessibility for visual and non-destructive examinations. The operating WCR plants have adopted proactive material ageing management programmes and state of the art ISI and degradation monitoring practices that will also be deployed for advanced WCRs. Therefore, a strong technical basis exists for developing current state of knowledge pipe failure rates that establish a good baseline for advanced WCR pipe failure rate estimation. Conclusions for continued research are provided in Section 12. Supporting information is presented in Annexes I–V, as follows:

— Annex I. Methodologies for Assessing Probabilistic Failure Metrics for Advanced WCRs;
— Annex II. Analysis Tools and Techniques;
— Annex III. Piping Reliability Data Resources;
— Annex IV. WCR Piping OPEX Summary;
— Annex V. Abstracts of Early Piping Reliability Analysis Studies.

2. LESSONS LEARNED FROM WCR OPERATING EXPERIENCE

2.1. PIPING RELIABILITY PRIMER

Pipe failures are a result of complex interrelated processes influenced by the design and construction characteristics, material properties, environmental conditions and external or internal loading conditions. Failures happen due to hydraulic and mechanical conditions imposing excessive loadings upon a degraded (e.g. cracked or thinned) piping component. The study of piping OPEX from operating WCRs yields technical insights into the interactions of many structural reliability characteristics and influence factors. Those insights are of high relevance in informing the development and application of piping reliability models.

Different types of piping reliability models have been developed. To varying degrees, all models rely on OPEX data and the statistics obtained from detailed evaluations of this data and experimental data. The quality, or validity, of analysis results rely on how well any relevant OPEX data have been acknowledged in the preparation of model inputs. For OPEX data to be as useful as feasibly possible, the failure event information is to be processed on the basis of a taxonomy on but not limited to the physics of material degradation and failure, fluid dynamics, fracture mechanics, piping system design and welding technology.

Developing a comprehensive pipe failure database represents therefore a multidisciplinary task that yields database structures consisting of on the order of 100 database fields needed to correctly characterize the filed experience data. The data classification needs many hundreds of key words (or data filters) in order to capture a multitude of possible combinations of piping reliability attributes and influence factors. The types of information found in a database on piping operating experience are illustrated in Fig. 2.

High level summaries of the current knowledge base on the WCR and advanced WCR piping OPEX are given in Subsections 2.2–2.14. The knowledge base which is available to analysts that pursue advanced WCR piping reliability assessments has significant advantages relative to the prior state of

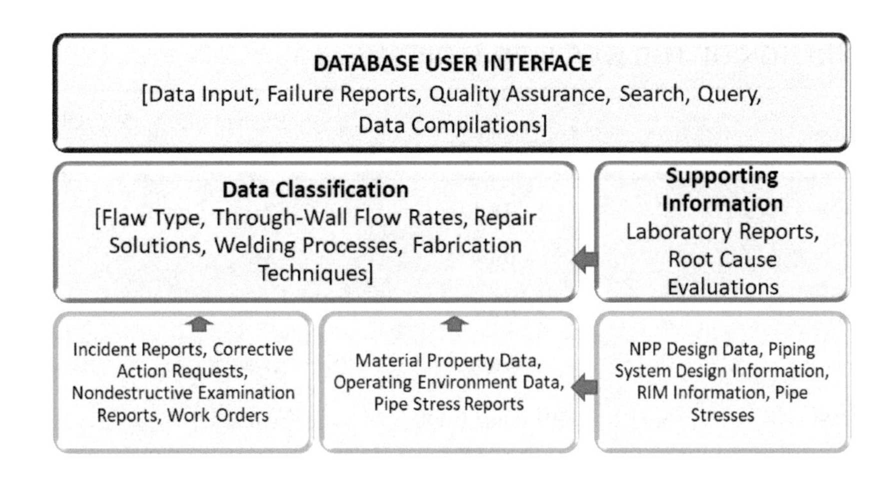

FIG. 2. Piping operating experience database structure and contents.

knowledge that was available to analysts when the first generation WCR technologies were deployed several decades ago. Also, the current methodologies and computational tools that are available to support the quantification of structural reliability parameters have evolved.

2.2. PIPING MATERIAL DEGRADATION

The material degradation mechanisms that have caused pipe failure are summarized in this subsection. Extensive piping material performance information exists, and it covers field experience that has been obtained over a period of about six decades[1] corresponding to more than 18 000 reactor-years of operation [28–34]. Examples of how the WCR piping OPEX has evolved are shown in Table 1 where the accumulated OPEX that was available in the early 1980s is compared with the current knowledge base.

TABLE 1. EVOLUTION OF THE WCR PIPING OPEX

Degradation mechanism	Plant type	S.H. Bush (1985)	OPEX as of 2021 (Annex II and IV)
Stress corrosion cracking	BWR	47.9%	17.3%
	PWR	6.4%	6.3%
Thermal fatigue	BWR	16.0%	0.9%
	PWR	7.7%	1.6%
High cycle fatigue	BWR	1.7%	6.4%
	PWR	7.1%	14.4%
Erosion-corrosion [a]	BWR	5,0%	1.8%
	PWR	1.4%	3.4%

[1] The first large scale WCR was Dresden Unit 1 (ca. 190 MWe) in the United States of America (USA). It was an early design BWR, which was first connected to the grid on 15 April 1960. The reactor was permanently shut down on 31 October 1978.

TABLE 1. EVOLUTION OF THE WCR PIPING OPEX (cont.)

Degradation mechanism	Plant type	S.H. Bush (1985)	OPEX as of 2021 (Annex II and IV)
Erosion-cavitation	BWR	0.0%	0.7%
	PWR	0.0%	1.6%
Flow accelerated corrosion	BWR	0.0%	6.5%
	PWR	0.0%	15.5%
Flow induced vibration	BWR and PWR	0.0%	0.7%
Corrosion-fatigue (environmental degradation)	BWR	1.1%	0.1%
	PWR	1.1%	0.4%
Corrosion (multiple mechanisms)	BWR	0.0%	3.4%
	PWR	0.0%	19.0%
Unknown cause	BWR	2.7%	0.0%
	PWR	1.9%	0.0%
Total no. events		350	11,500

[a] Reference [35] includes a single line item for 'erosion-corrosion' in lieu of flow assisted degradation, which includes four different types of wall thinning mechanisms; erosion-corrosion, erosion-cavitation, flow assisted corrosion and flow induced vibration.

The interactions of applied mechanical stresses, operating environments (such as corrosion potential, flow conditions, pressure, temperature, humidity) and metallurgy (such as chemical and mechanical material properties) cause degradations and failures of piping material. The collected knowledge based on the operating WCRs points to the following categories of material degradation mechanisms and stressors (shown in Fig. 3):

— Corrosion fatigue, also known as environmental degradation.
— Corrosion attack, including microbiologically influenced corrosion, pitting, crevice corrosion, corrosion under insulation, galvanic corrosion and graphitic corrosion.
— Design and construction defects representing a broad category that includes weld defects (or lack of fusion), inadequate or lack of coating, inadequate or lack of pipe supports, inadequate cathodic protection, inadequate heat tracing.
— Flow assisted material degradation includes erosion, erosion-cavitation, erosion-corrosion, flow accelerated corrosion and flow induced vibration or abrasive/fretting wear.
— Hydrogen embrittlement or hydrogen assisted cracking from long term exposure to high temperature and elevated hydrogen levels.
— High cycle and low cycle fatigue.
— Severe overloading caused by external impact (e.g. accidental heavy load drop on piping) or internal over pressurization through a hydraulic transient or hydrogen deflagration.

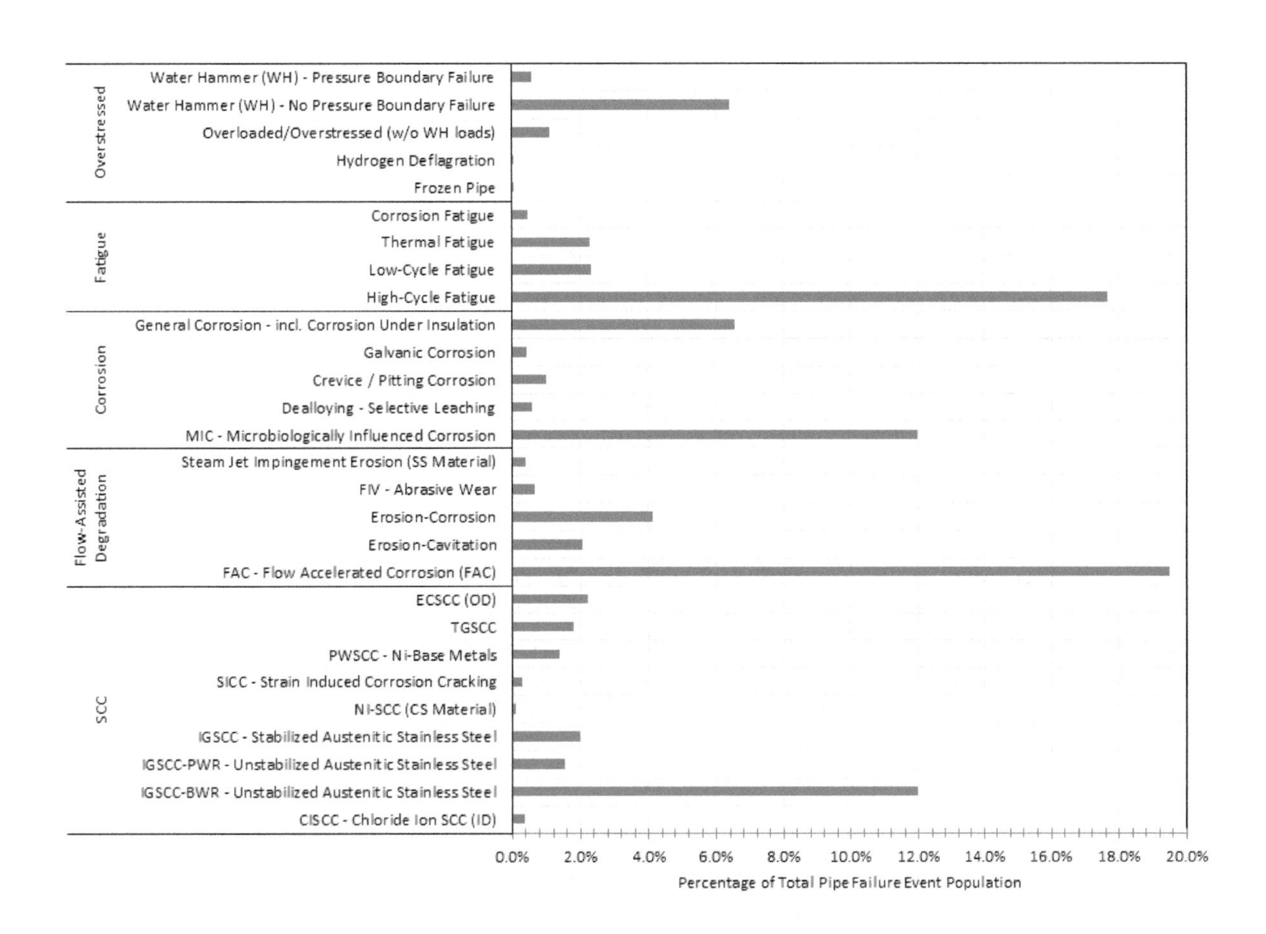

FIG. 3. WCR piping material degradation mechanisms.

— Stress corrosion cracking (SCC), including intergranular and transgranular SCC. SCC mainly affects stainless steels, and nickel base materials given certain environmental and stress conditions.
— Thermal fatigue involves thermal stratification, thermal cycling, thermal striping and thermal sleeve mixing.

2.3. PIPE FAILURE MANIFESTATIONS

The causes of piping degradations and failures are due to various damage or degradation mechanisms. Piping failure occurs due to synergistic effects involving operating environments and loading conditions. The WCR piping OPEX that is summarized in Fig. 3 is explored in further detail in Fig. 4; Annex III includes additional background information. There are two forms of pipe failure:

— *Event driven failure.* Mechanically stress driven, and is attributed to conditions involving combinations of equipment failures (other than the piping itself, e.g. loose/failed pipe support, leaking valve) and stress risers or unanticipated loading conditions (e.g. hydraulic transient or operator error that causes an inadvertent valve operation). Short term degradation is sometimes used in lieu of event-driven degradation [36].
— *Failure attributed to time dependent environmental degradation.* Defined by the conjoint requirements that include operating environment, material and loading conditions. These conjoint requirements differ across the different types of piping system designs and are influenced by routeing, material,

diameter, wall thickness, method of construction/fabrications, etc. Similarly, pipe flaw incubation times and flaw growth rates differ across the many combinations of degradation susceptibility and operating environment. Non-destructive examination techniques usually detect an onset of material degradation before a crack propagates through a pipe wall.

A failure, referred to in Fig. 4, is any degraded condition with an operational impact and results in pipe repair or replacement. Determined by non-destructive examination techniques or metallographic examination, non-through-wall defects are characterized by the depth of a crack relative to the pipe wall thickness, orientation and length of a crack. National codes and standards define what is acceptable and not acceptable for continued operation. Through-wall defects are characterized by the size of a pipe wall penetrating flaw and the resulting mass or volumetric leak or flow rate. The downward arrows in Fig. 4 symbolize the potential synergistic effects of various damage and degradation mechanisms. As one example, various types of weld defects (e.g. lack of fusion, slag inclusions) tend to be strong contributors to crack initiation sites that ultimately result in an SCC failure. As another example, thermal fatigue can cause crack initiation while an SCC mechanism can cause crack propagation in a pipe through-wall direction. The vertical arrows indicate the presence of synergistic effects. For example, thermal fatigue may cause crack initiation, and crack propagation may occur via intergranular stress SCC. As another example, a pre-existing weld defect could be the source of an SCC sequence. The fill-effects in the coloured horizontal bars are commensurate with the observed event populations (i.e. a strong fill corresponds to multiple events and a weak fill corresponds to a few major structural failures).

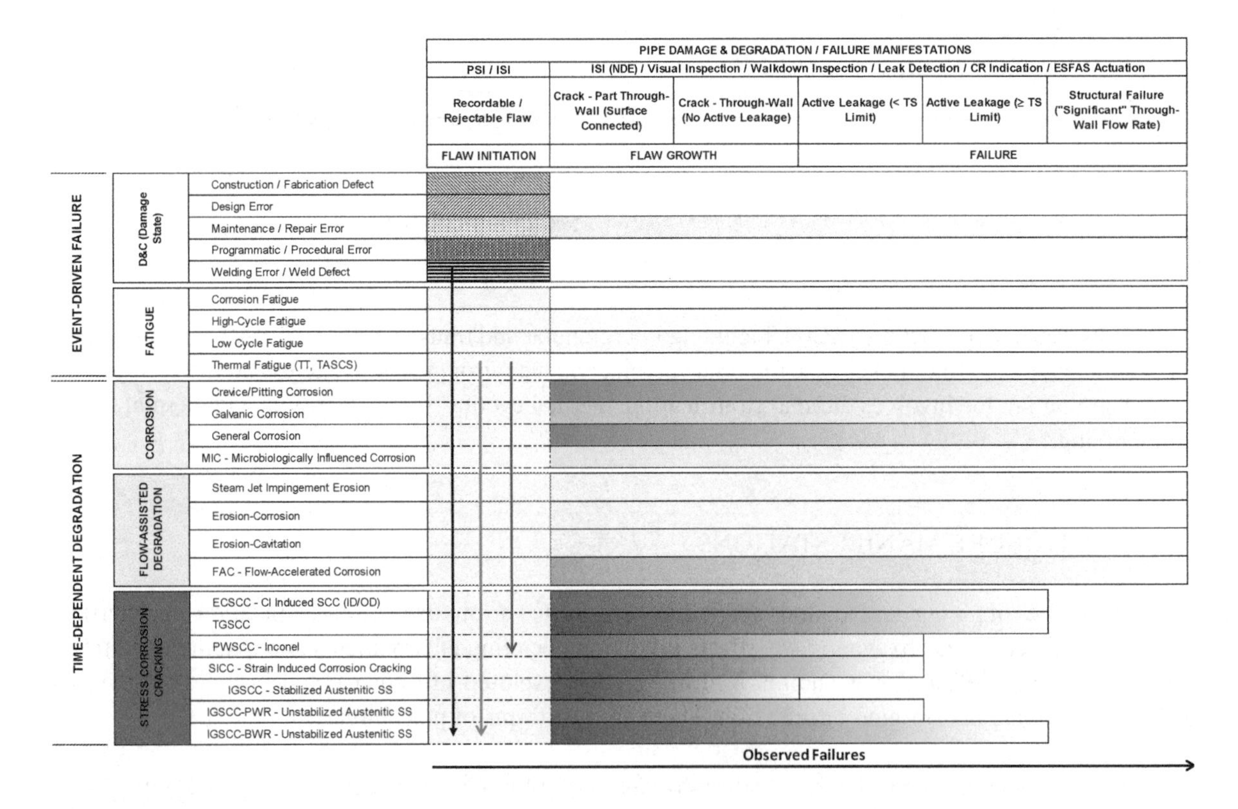

FIG. 4. Pipe failure manifestations. The vertical arrows indicate the presence of synergistic effects. For example, thermal fatigue may cause crack initiation, and crack propagation may occur via intergranular stress SCC. As another example, a pre-existing weld defect could be the source of an SCC sequence. The fill-effects in the coloured horizontal bars are commensurate with the observed event populations (i.e. a strong fill corresponds to 'multiple' events and a weak fill corresponds to a few major structural failures).

Certain combinations of metals/environment systems, localized loading or stress conditions, and methods of fabrication/installation have produced major structural failures while other combinations at most have resulted in relatively minor through-wall flaws. For example, stainless steel piping in primary water environments has not experienced any major structural failures caused by environmental degradation. On the other hand, carbon steel in wet steam, high temperature environments has experienced major structural failures.

2.4. CORROSION MECHANISMS

Corrosion is a naturally occurring phenomenon defined as the deterioration of a metal that results from a chemical or electrochemical reaction with its environment. Corrosion takes many different forms and is affected by numerous environmental factors. Three types of corrosion mechanisms are shown in Fig. 5.

Process piping (e.g. auxiliary cooling water systems) in a raw water environment is susceptible to microbiologically influenced corrosion produced by local environments. There are various types (modes) of such corrosion: general corrosion, pitting, crevice corrosion, de-alloying, galvanic corrosion, intergranular corrosion, SCC and corrosion fatigue. The growth of microbes is promoted by slowly flowing environments in providing a ready supply of nutrients (in particular, periodically flushed systems). Stagnant areas adjacent to flowing areas are particularly susceptible to microbiologically influenced corrosion.

Microbiologically influenced corrosion is relatively common in low temperature systems such as, but not limited to, fire water, service water and circulating water systems. It typically occurs in two general locations: on external surfaces where there is moisture and other materials, such as organic debris buildup, which contains nutrients suitable for bacterial or fungal growth; and on internal surfaces in low temperature components — primarily where water is flowing slowly or is periodically flushed. Both conditions provide the required supply of nutrients for microbiological activity and growth. Corrosion of carbon and low alloy steel components by leaking borated water has caused significant problems for many pressurized water reactor (PWR) plants. Of the greatest concern is uniform corrosion, referred to as wastage.

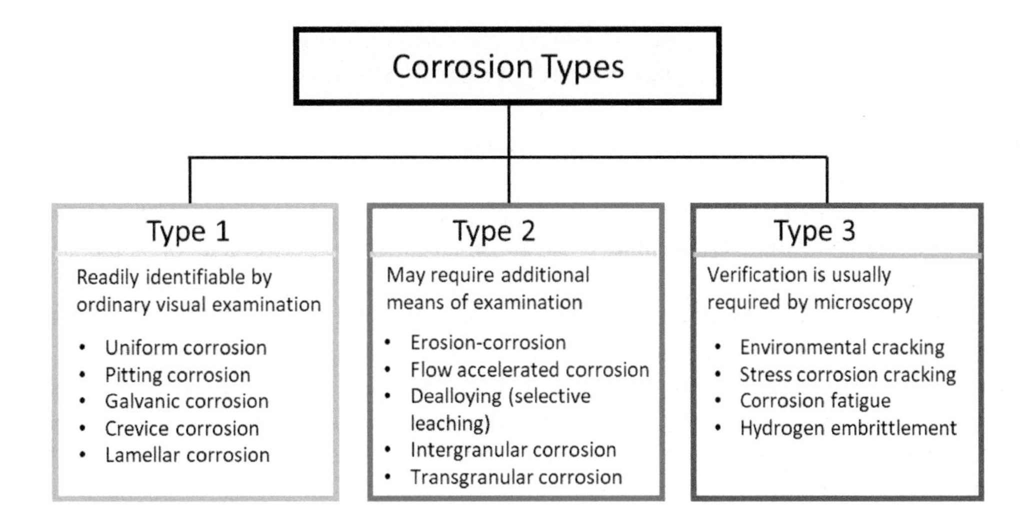

FIG. 5. The different types of corrosion mechanisms.

2.5. DESIGN AND CONSTRUCTION DEFECTS

Insights from root cause analyses of pipe failures point to the significance of human error (or organizational factor) contributions. Process industry incident statistics show that 20–90% of all incidents are indirectly or directly caused by human error. Human errors are either latent or active. Effects of a latent error may lie dormant within a system for a long time, only becoming evident after a period of time when the condition caused by the error combines with other errors or particular operating conditions. An example of latent error affecting piping reliability is the design or construction error first revealed, say, several years after commercial operation began. Another example of latent human error affecting piping reliability is an ISI programme that does not fully acknowledge relevant operating experience with a particular type of piping system. By contrast, effects of an active human error are felt almost immediately (e.g. water hammer due to improper post-maintenance restoration of a piping system).

Many studies have been performed to assess the human error contributions to pipe failure. Hurst et al [37] analysed pipe failures in the chemical process industry and showed that 'operating error' was the largest immediate contributor to piping failure (30.9% of all known causes). Overpressure (20.5%) and corrosion (15.6%) were the next largest categories of known immediate causes. The other major areas of human contribution to immediate causes were human initiated impact (5.6%) and incorrect installation of equipment (4.5%). The total human contribution to immediate causes was therefore about 41%. For the underlying causes of pipe failure, maintenance (38.7%) and design (26.7%) were the largest contributors. The largest potential preventive mechanisms were human factors review (29.5%), hazard study (25.4%) and checking and testing of completed tasks (24.4%). A key conclusion of the study was that about 90% of all failure events could have been prevented through adequate RIM processes (e.g. ISI, leak detection, application of mitigation processes to reduce or eliminate the potential for material degradation).

2.5.1. Design and construction defects event sequence diagram

Service induced material degradation is a result of synergies among material properties, loading (e.g. stress riser) and environmental conditions. Through-wall pipe flaws involve initiation and incubation. A pre-existing flaw can act as a stress riser causing a crack initiation and progress in a through-wall direction if exposed to an adverse environment. The majority of pipe flaws that result in a corrective action (repair or replacement) are attributed to a readily identifiable active degradation mechanism or an off normal loading condition.

A relatively small subset of all recordable or rejectable pipe failures involves a pre-existing defect that grows over a long time and is detected through a surface examination (e.g. visual examination or liquid penetrant testing). The design and construction defects event sequence diagram (Fig. 6) shows an example of how to classify weld flaws for which no active degradation mechanism is present. As an example, safety class 1 welds are subjected to pre-service inspection and rejectable flaws are repaired. There is some likelihood that a pre-existing flaw is not detected, however, an ISI may/may not detect a weld defect. If successfully detected, the weld defect is evaluated per ISI programme acceptance standards. Continued operation is possible if repair/replacement is performed, or some degradation mitigation is implemented.

Under the assumption that a pre-existing flaw is discovered during an ISI but remains unmitigated (i.e. no repair is performed) then crack growth may occur given that it is subjected to an adverse operating environment (high temperature, corrosive, subject to high stresses) and the material is susceptible to degradation. The potential of through-wall cracking would be high if the conjoint requirements for degradation are met (through-wall crack potential H in Fig. 6). In developing an initiating event frequency model, pipe failure rates and rupture probabilities are derived for all piping components within the evaluation boundaries.

A pipe failure rate estimation process considers all credible damage and degradation mechanisms that apply to an evaluation boundary. For locations without any readily identifiable damage or degradation susceptibility an assumption is made that a pre-existing weld flaw may exist and eventually grow in the

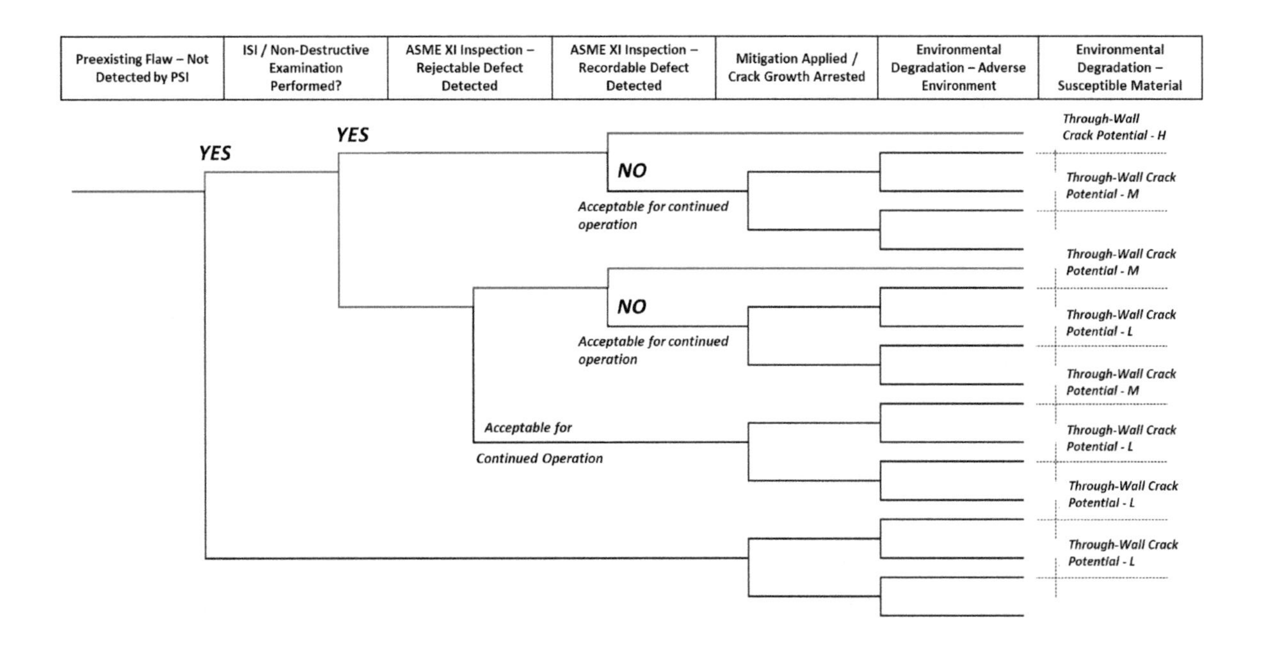

FIG. 6. D&C event sequence diagram.

through-wall direction. The flaw growth mechanism is termed 'low cycle fatigue and pressure loading' and accounts for the effects of normal operation including cooldown and heat-up cycles. Therefore, a conditional failure probability (CFP) model is needed to resolve the low cycle fatigue analysis cases. Highlighted in red, the two event sequences in Fig. 6 represent a high likelihood that a pre-existing defect will eventually grow and penetrate the outside pipe wall.

2.5.2. Effect of welding processes on the material degradation susceptibility

The mechanical properties and metallurgical characteristics of pipe welds are affected by welding processes such as type and speed of welding. Differential thermal expansion and contraction of the weld metal and parent material cause welding residual stresses. Fatigue and SCC require the presence of tensile weld residual stresses. Crack initiation occurs with enough load cycles at a high enough tensile stress. Some qualitative insights can be drawn from the WCR OPEX regarding the effect of welding processes on material degradation. The OPEX data distinguish field weld failures from shop weld failures. This leads to the question of whether field welds are less or as reliable as shop welds. It is a complicated task to assess the many factors that affect weld quality and failure propensity.

Solution heat treatment is used for shop welds. It reduces or eliminates weld sensitization and residual stresses. Figures 7 and 8 illustrate the experience with field welds and shop welds. In boiling water reactor (BWR) plants with external reactor recirculation loops, approximately 40% of the welds are shop welded. Except for pipe to safe-end welds (or terminal welds), all safety class 1 PWR reactor coolant system (RCS) hot leg, cold leg and surge line welds are shop welded.

Investigations into intergranular SCC of main coolant circuit welds in high power channel type nuclear reactor (RBMK or Reaktor Bolshoy Moshchnosty Kanalny) plants have pointed to a significantly higher incident rate for welds in vertical pipe runs than horizontal pipe runs (Fig. 8). The reason given is that per the welding procedure specification, at the time of construction, higher heat input was required than for horizontal welds [38]. The higher heat input caused higher sensitization of the heat affected zones and higher weld residual stresses.

The OPEX data also identify the location of a failed weld in a pipe spool (e.g. pipe to pipe, elbow to pipe, safe-end to pipe). Figure 9 shows the location dependency of weld failures and reflects how different pipe stresses affect the integrity.

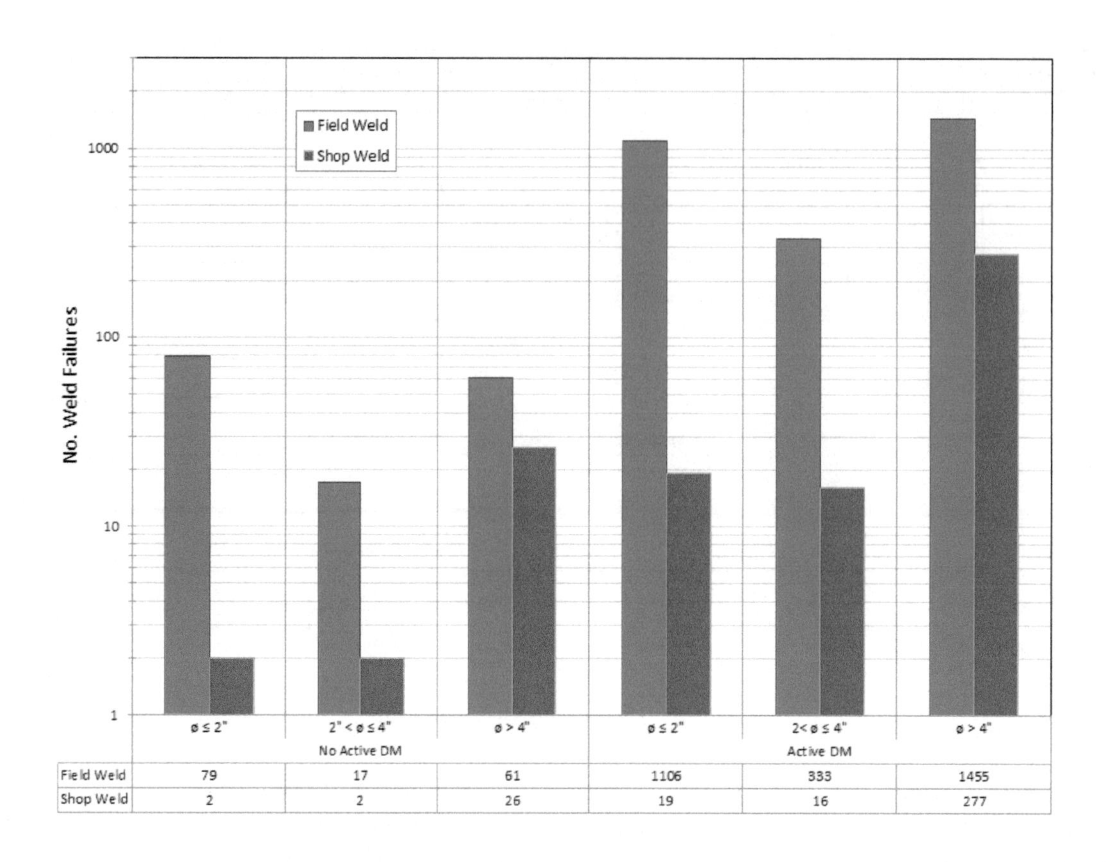

FIG. 7. BWR primary system field and shop weld OPEX.

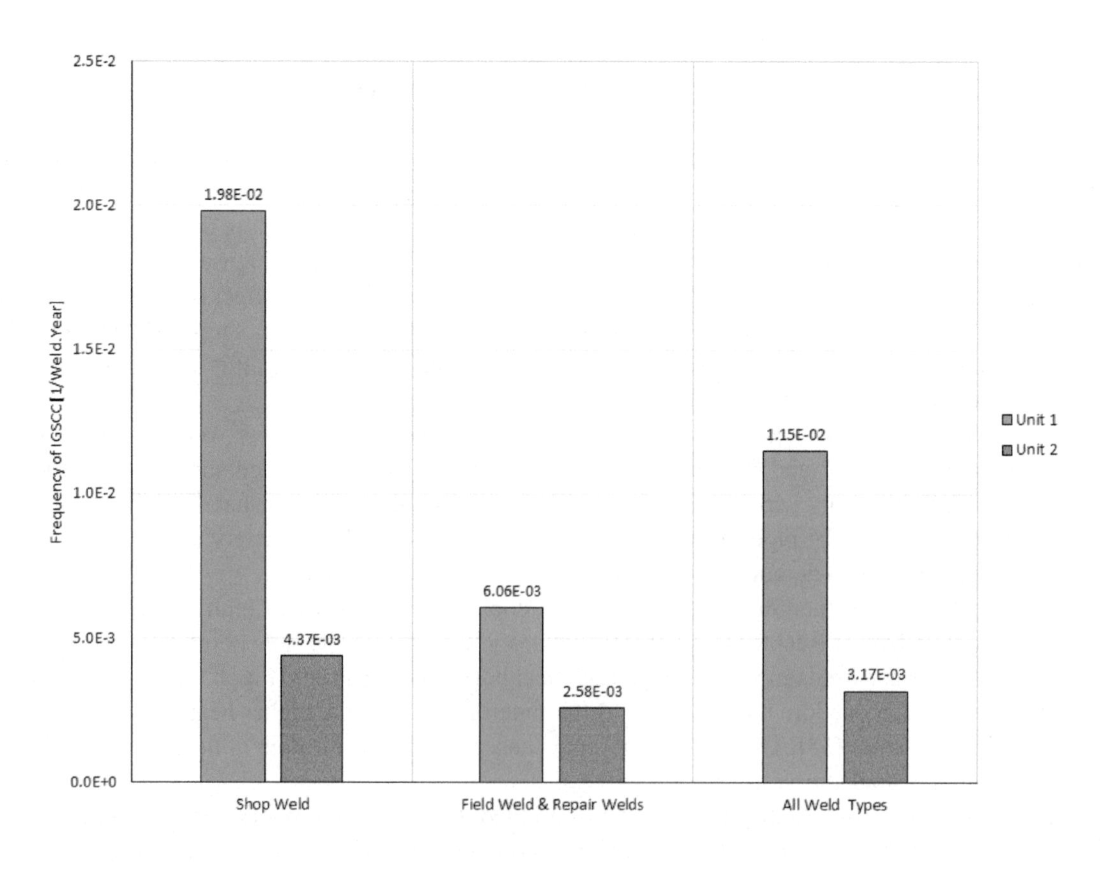

FIG. 8. RBMK primary system field and shop weld OPEX.

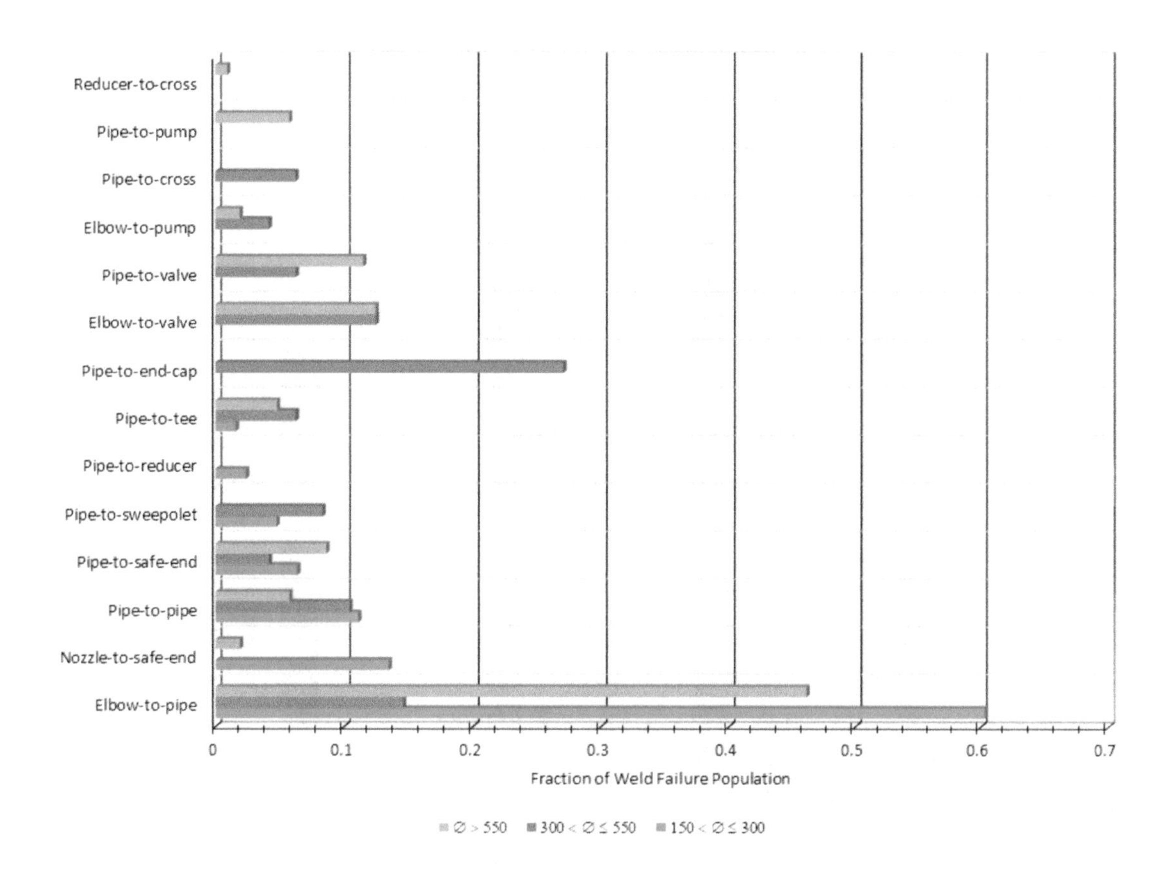

FIG. 9. Weld OPEX as a function of in-line location and pipe size.

2.6. FATIGUE OF PIPING COMPONENTS

Fatigue is the process of progressive localized permanent structural change occurring in material subjected to conditions which produce fluctuating stresses and strains at some point or points and which may culminate in crack or complete fracture after a sufficient number of fluctuations. There are four basic types of fatigue mechanisms: (1) corrosion fatigue (Section 2.7), (2) low cycle fatigue, (3) thermal fatigue, and (4) high cycle fatigue. Figure 10 summarizes the WCR OPEX involving pipe failures caused by fatigue mechanisms. Approximately 80% of WCR fatigue failures are due to high cycle vibration fatigue of small diameter butt welds and socket welds.

The term 'low cycle fatigue' is used to characterize crack growth in the pipe through-wall direction through applied stress and normally occurring cooldown/heat-up cycles. An underlying assumption is that of a pre-existing weld flaw attributed to original construction, fabrication or welding defects missed by pre-service inspections and/or subsequent ISIs. The following conditions are to be met for an event to be classified as low cycle fatigue: no active environmental degradation mechanism can be identified, and the root cause evaluation points to the presence of a weld flaw such as lack of fusion.

The cyclic stresses resulting from changing temperature in a component or in the piping attached to the component causes thermal fatigue; it may include a relatively low number of cycles at a higher strain (such as plant operational cycles or injection of cold water into a hot nozzle) or due to a high number of cycles at low stress amplitude (as local leakage effects or cyclic stratification). There are several different thermal fatigue phenomena including: thermal stratification, thermal cycling, thermal striping, valve in-/out-leakage and thermal sleeve mixing.

High cycle fatigue involves a high number of cycles at relatively low stress amplitudes (typically below the material's yield strength but above the fatigue endurance limit of the material). The crack

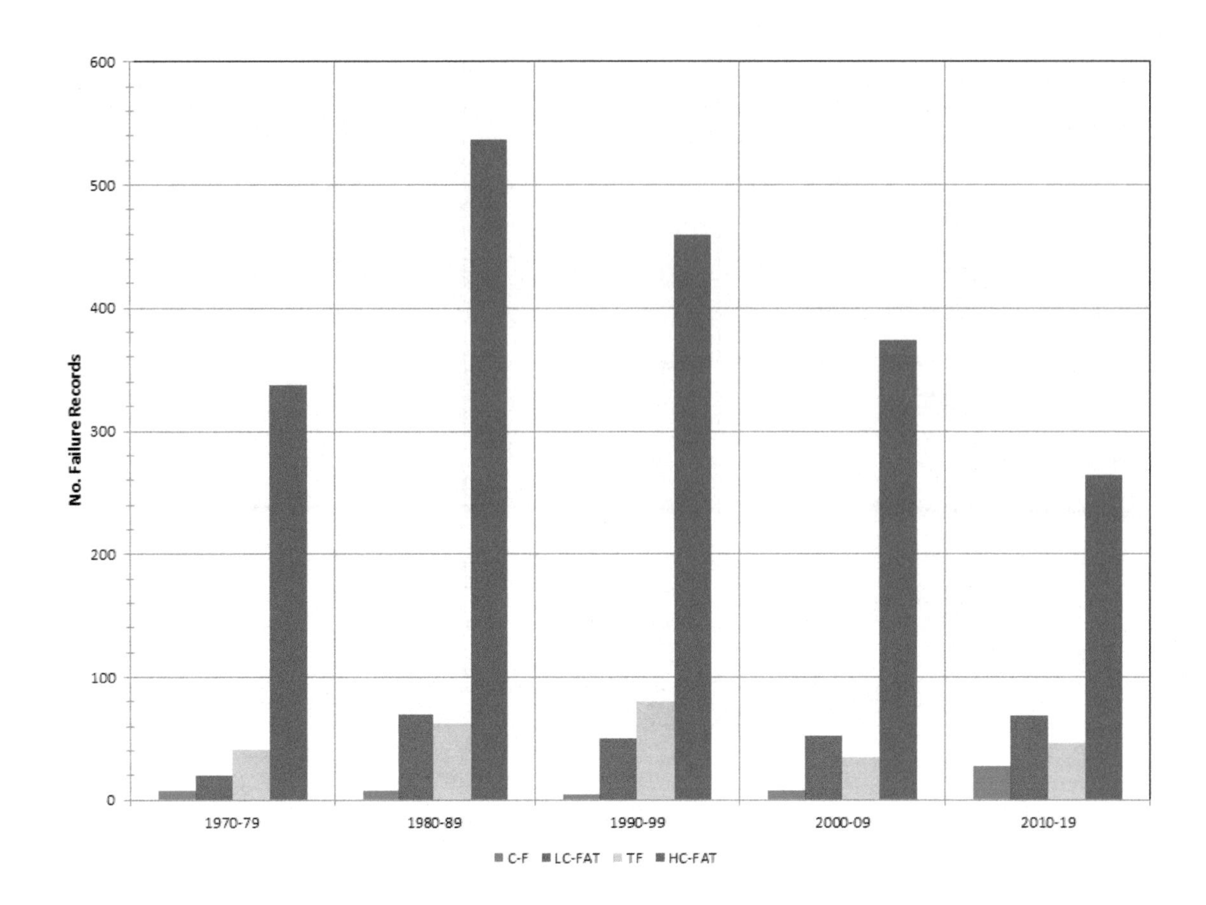

FIG. 10. WCR pipe failures caused by fatigue mechanisms.

initiation phase is considered to be dominant, since crack growth is usually fairly rapid. High cycle fatigue may be due to vibration or pressure pulses or due to flow induced vibration.

2.6.1. Contributing factors to high cycle fatigue

The following description follows details provided in [32]. Fatigue in piping can be induced by inadequate piping supports. The natural frequency of a piping system is influenced by the type and distance of piping supports such that tightly spaced restraints increase the piping system's frequency while widely spaced restraints reduce it. Thus, high frequency fatigue damage may result when an excitation source exists in the natural frequency range of the piping system causing the resonance of a piping system to occur. The main contributing factors are mechanical excitation mechanisms, cavitation excitation and flashing.

An example of mechanical excitation mechanism is the pressure pulsations in centrifugal pumps and positive displacement pumps occurring at frequencies that are multiples of the vane passing frequency. They occur in the range of acoustic frequencies and are continuous and propagate through the coolant medium just as sound is transmitted through air. Acoustically induced vibrations can be defined as vibrations induced by acoustic pressure waves generated by a noise source in the reactor, by a high speed or unsteady gas flow, or by the interaction of gas flows and solid objects. These pressure waves propagate in the fluid system with the speed of sound and are strong enough to cause unacceptable vibrations [39–41]. If pressure pulsations happen to coincide with the structural frequency of the piping system, severe vibratory fatigue damage may take place in the system.

The fluid pressure approaching its vapour pressure may cause cavitation excitation. Under such conditions, small fluctuations may induce the formation of vapour pockets, and when they rapidly collapse they generate intense shock waves. The resulting pressure pulsation can cause severe piping vibration downstream of the component. Adding to severe vibrations, the collapse of the cavities impacts the solid surface and removes material by mechanical erosion.

Flashing occurs when the coolant temperature becomes higher than its saturation temperature at a given pressure and the coolant flashes into steam. This results in broadband pressure pulsations, causing vibration of the piping downstream of the flashing component and results in steam hammer phenomena. Because of this phenomenon, the control valves that are regulating the flow of oversaturated coolant into a vessel kept at lower pressure are located as close as possible to the vessel entry nozzle; this way the flashing may occur in the larger volume of the vessel and not in the piping.

2.6.2. Fatigue induced electro-hydraulic control piping failures

The electro-hydraulic control system is a non-safety system consisting of small diameter piping at high operating pressure; between 7 MPa and 15 MPa. The system is vital for power generation. A failure of this system can result in turbine/reactor trip or in preventing the startup of a reactor. Turbine trip followed by reactor trip due to electro-hydraulic control circumferential pipe break is not an uncommon occurrence [42]. A failure could also result in a main turbine overspeed condition with a turbine blade and/or rotor failure that generates missile hazards that could damage nearby safety related systems. The circulating turbine fluid used is triaryl phosphate esters, which provides improved fire resistance over hydrocarbon based fluids, and thereby minimizes the risk of fire due to a high pressure electro-hydraulic control fluid leaking on to a hot surface or other ignition source.

The electro-hydraulic control fluid system consists of numerous runs of small diameter hydraulic piping and tubing to convey its hydraulic fluid to the valve actuators where it is needed. The piping material is typically type 304 stainless steel for the tubing and piping and type 316 stainless steel for the fittings.

2.6.3. Socket weld integrity management

There continues to be frequent occurrences of high cycle fatigue failures of socket welded connections in safety related piping systems. The use of socket welds in safety class 1 systems varies extensively across WCR plants: from about 50 to 500 welds in a WCR.

There are significant country to country differences in codes and standards for the use of socket welds in safety related piping. As an example, in 2002 the French Nuclear Safety Authority issued a directive concerning socket weld integrity. According to this directive, socket welds not meeting the requirements for weld dimensions and/or weld integrity as specified by the Rules for the Design and Construction of Mechanical Equipment for Nuclear Power Plants PWR Code are to be replaced with butt welds [43].

According to the US Nuclear Regulatory Commission (NRC), for a one-time inspection to detect cracking in socket welds, the inspection is to be either a volumetric or opportunistic destructive examination that is performed when a weld is removed from service for other considerations (e.g. plant modifications). In the case when more than one weld is removed, a sampling basis is used. These examinations provide additional assurance that either ageing of small bore ASME code class 1 piping is not occurring or the ageing is insignificant, such that a plant specific ageing management programme is not warranted and is applicable to small bore ASME code class 1 piping and systems less than DN100 and greater than or equal to DN25.

2.7. CORROSION FATIGUE

Corrosion fatigue or environmentally assisted fatigue is the behaviour of materials under cyclic loading conditions and in a corrosive, high temperature/high pressure operating environment. It is considered to be made up of a region (or life) associated with the formation of an engineering sized crack, and a region consisting of the growth of this crack, up to component failure. One category relates to the cycling life for the formation of a fatigue crack in a smooth test specimen, the so-called *S-N* fatigue properties (stress versus number of cycles). The second relates to the growth of a pre-existing crack. Laboratory tests have shown that coolant water in light water reactors (LWRs) can have a detrimental effect on both *S-N* fatigue properties and fatigue crack growth. Much lower failure stresses and much shorter failure times can occur in a corrosive environment compared to the situation where the alternating stress is in a non-corrosive environment.

Corrosion fatigue is not to be confused with stress corrosion, which is crack initiation and growth under sustained load or residual stress. Corrosion fatigue is a mostly transgranular crack growth phenomenon. The corrosion fatigue fracture is brittle and the transgranular cracks are not branched. The corrosive environment can cause a faster crack growth and/or crack growth at a lower tension level than in dry air. Even relatively mild corrosive atmospheres can reduce the fatigue strength of aluminium structures considerably, down to 75–25% of the fatigue strength in dry air. No metal is immune from some reduction of its resistance to cyclic stressing if the metal is in a corrosive environment. Control of corrosion fatigue can be accomplished by either lowering the cyclic stresses or by various corrosion control measures.

Results from laboratory tests generally reveal a detrimental effect of WCR coolant environments on the fatigue lives of specimens made from carbon steels, low alloy steels, austenitic stainless steels and nickel base alloys. The parameters predominantly affecting the fatigue life of laboratory specimens are strain rate, temperature, dissolved oxygen concentration in the water and sulphur content of the material, the latter of which is only applicable for carbon steels and low alloy steels.

The detrimental effects of reactor environments on fatigue lives have been known for more than 30 years. Reactor coolant pressure boundary components exposed to the reactor water environment have exhibited degradation due to environmentally enhanced fatigue in service. In all these cases, unacceptable component fabrication, material selection, or plant operation (and combinations of these) were identified as root causes leading to the degradation. Significant large scale, generic degradation due to environmental fatigue has not been observed in service even though environmental effects due to the impact of WCR coolant were not explicitly considered in current design rules. The US NRC investigation of the risk associated with corrosion fatigue in the Fatigue Action Plan SECY–95–245 issued in 1995[2] concluded that there was no inherent risk to core damage frequency for operating nuclear reactors, although increased probability of leakage indicates that this issue requires management for extended plant operation.

Limited observations of cracking due to corrosion fatigue stand in contrast to significant occurrences of SCC in stainless steels and nickel base alloys, which have been observed more systematically in reactor coolant pressure boundary welds and reactor internals from LWR plant operational experience worldwide.

The lack of significant observed degradation in plant components with regard to corrosion fatigue is attributed, at least in part, to the generally conservative design requirements adopted within the ASME code and applicable regulations to keep the cumulative usage factor <0.1 for break exclusion locations. Margins in the design requirements appear to compensate for the detrimental environmental effects.

Another consideration when comparing the environmental effects between laboratory and service components is the applied loading associated with pressure and thermal transients. Laboratory testing typically relies on simple mechanically controlled loading transients (e.g. artificially shaped waves) and may include some amount of compensation for the effects of more complex thermal transient loading. Additionally, plant components are often subjected to thermal transients with long-lasting hold times at almost constant load or temperature corresponding to steady state operating conditions, which may lead to some strain recovery within the component. These differences may affect fatigue lives.

[2] See: US NUCLEAR REGULARLY COMMISSION, Policy Issue, Completion of the Fatigue Action Plan (1995), https://www.nrc.gov/docs/ML0314/ML031480210.pdf

2.8. FLOW ASSISTED DEGRADATION

The term 'flow assisted degradation' encompasses several phenomena, all of which result in the degradation of piping through material loss. These phenomena include erosion, erosion-cavitation, erosion-corrosion, flow accelerated corrosion and liquid droplet impingement erosion. Historically, the erosion-corrosion and flow accelerated corrosion terms were used interchangeably to describe similar material degradation processes. Both types of damage involve destruction of a protective oxide film on the inside pipe wall. The removal of the oxide film is generally referred to as the erosion process. This is followed by electrochemical oxidation, or corrosive attack of the underlying metal. The differences between erosion-corrosion and flow accelerated corrosion involve the mechanism by which the protective film is removed from the metal surface. In the erosion-corrosion process the film is removed mechanically from the surface. In contrast, in the flow accelerated corrosion process the oxide is dissolved or prevented from forming, allowing corrosion of the unprotected metal. Flow accelerated corrosion occurs in two-phase flow conditions (e.g. water droplets in steam or steam bubbles in water) as well as single-phase flow conditions. The main distinguishing characteristics of the different flow assisted degradation mechanisms are summarized in Fig. 11.

2.8.1. Erosion-cavitation

Erosion-cavitation is the process of surface deterioration and surface material loss due to the generation of vapour or gas pockets inside the flow of liquid [44]. These pockets are formed due to low pressure well below the saturation vapour pressure of the liquid and erosion caused by the bombardment of vapour bubbles on the surface. Erosion-cavitation usually involves an attack on the surface by gas or vapour bubbles, creating a sudden collapse due to a change in pressure near the surface. Low pressure (below the saturated vapour pressure) is generated hydrodynamically due to various flow parameters, such as liquid viscosity, temperature, pressure and nature of flow. This deterioration is initiated by a sudden surge of bubbles hammering the surface, resulting in deformation, as well as pitting.

2.8.2. Erosion-corrosion and liquid droplet impingement erosion

Erosion-corrosion is a mechanism of material loss by mechanical means due to impingement, abrasion or impact, etc., resulting from the movement of a liquid or gas over the surface of a metal coupled with corrosion. This type of degradation is characterized by attack like small pits with bright surfaces free

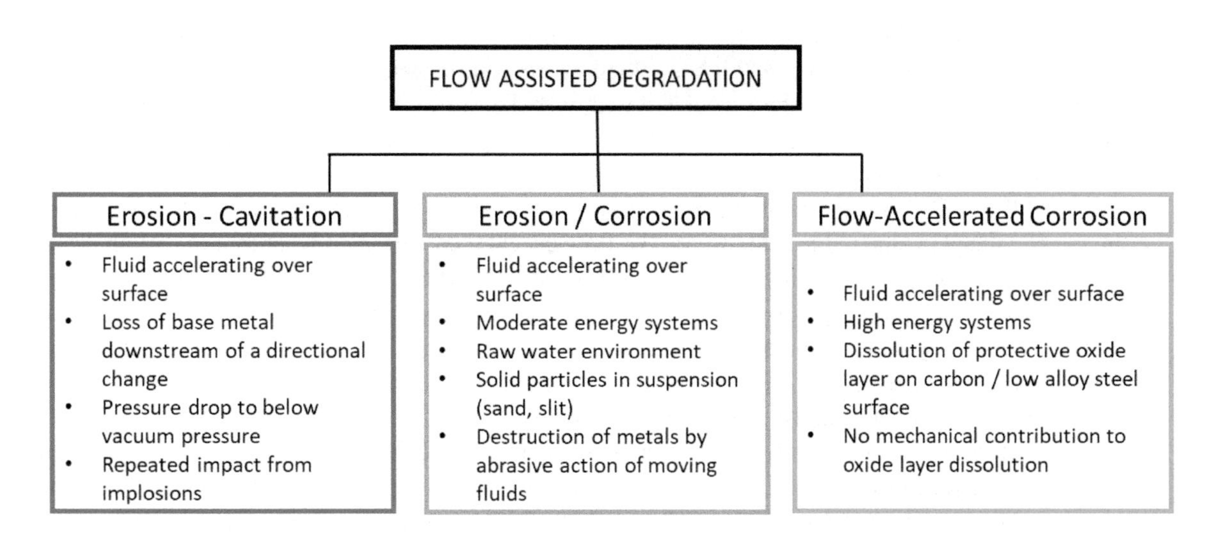

FIG. 11. Distinguishing characteristics of the different flow assisted degradation mechanisms.

from corrosion products. These pits often have the form of a horseshoe with the nib pointing in the current direction. Erosion-corrosion may occur where the velocity of liquid is too high. Most exposed are places where there are effects of turbulence (e.g. joints, bends, etc.). The corrosion rate will accelerate if the liquid contains gas bubbles and/or solid particles. Systems susceptible to erosion-corrosion include raw water cooling systems (e.g. circulating water and service water systems).

The liquid droplet impingement erosion is a subset of erosion-corrosion. Liquid droplets are often generated in piping that operates in a two-phase flow condition and is due to the entrainment of liquid water from the upstream and also by the heat transfer through the pipe wall. In the region behind the orifice and the valve in the pipeline, the velocity of the droplets is highly accelerated due to the contraction effect. This results in the occurrence of high impact pressure on the inner surface of the pipe due to the liquid droplet impingement. The impact pressure of the droplets increases as high as several hundred MPa, which is beyond the elastic limit of the pipe wall material, so that the pipeline is often damaged by the impact pressure of droplets. In general, the liquid droplet impingement occurs on the dorsal side (extrados) of a bend or elbow, where the droplets cannot follow the steam flow due to the inertia of the droplets.

2.8.3. Flow accelerated corrosion

Flow accelerated corrosion leads to wall thinning (or metal loss) of steel piping exposed to flowing water or wet steam. The wall thinning is the result of the dissolution of the normally protective oxide layer formed on the surfaces of carbon and low alloy steel piping. The rate of metal loss depends on a complex interplay of several parameters including water chemistry, material composition and hydrodynamics, but based on OPEX the metal loss can be as high as 3 mm/yr. Carbon steel piping components that carry wet steam are especially susceptible to flow accelerated corrosion and represent an industry wide problem. The major parameters and factors affecting flow accelerated corrosion are [43]:

— *Effect of temperature.* An important variable affecting the flow accelerated corrosion resistance of carbon and low alloy steels is temperature. Most of the reported cases of flow accelerated corrosion damage under single-phase conditions have occurred within the temperature range of 80–230°C, whereas the range is displaced to higher temperatures (140–260°C) under two-phase flow. The exact location of the maximum wear rate changes with pH, oxygen content and other environmental variables. Experience has shown that the wear rate is highest at around 150°C and increases with fluid velocity. Furthermore, flow accelerated corrosion can occur in low temperature single-phase systems under unusual and severe operating conditions.
— *Effect of flow velocity.* Flow rate of the liquid has been found to have a linear effect on the flow accelerated corrosion wear rate. As higher velocities are experienced, higher wear rates are expected. Since the enhanced mass transfer associated with turbulent flows is the fundamental process in the accelerated dissolution of the pipe wall protective oxide layer, the effect of flow is best described in terms of the mass transfer coefficient.
— *Effect of fluid pH.* Flow accelerated corrosion wear rates are strongly dependent on pH. In general, increasing the pH value reduces the wear. The flow accelerated corrosion wear rate of carbon steels increases rapidly in the pH range of 7–9 and drops sharply above pH = 9.2. As the fluid becomes more acidic, more pipe wall losses are expected. The pH value can be affected by the choice of control agents (e.g. morpholine or ammonia) and by impurities in the water. In two-phase flows the critical parameter is the pH of the liquid phase. This can be significantly affected by the partitioning of the control agent between the steam and liquid phase.
— *Effect of oxygen.* Flow accelerated corrosion rates are inversely affected by the amount of dissolved oxygen in the feedwater, and too low an oxygen level is harmful to carbon steel piping. The flow accelerated corrosion rate decreases rapidly when the water contains more than 20 ppb oxygen, but the precise oxygen level required to prevent flow accelerated corrosion depends on other factors such as pH and the presence of contaminants. In BWRs, hydrogen water chemistry can be applied

with the main intention to suppress intergranular SCC susceptibility and crack growth rate. The flow accelerated corrosion rate has been measured in a laboratory test to be higher for a time period of eight months after starting hydrogen water chemistry. After this time, the flow accelerated corrosion rate appears to be similar to that in a reference normal water chemistry environment. Certain guidelines consider an oxygen level of 20–50 ppb desirable for hydrogen water chemistry. Some plants do add oxygen in their feedwater when using hydrogen water chemistry, while others do not. The use of noble metals to reduce the quantities of hydrogen required to establish hydrogen water chemistry conditions has, to date, not had a more pronounced effect on flow accelerated corrosion than the application of hydrogen water chemistry itself. Main steam lines made of carbon steel are susceptible to flow accelerated corrosion in the steam phase because most of the oxygen, being a gas, remains in the steam phase and does not partition to the liquid. For the same reason, injection of oxygen into the wet steam will not prevent flow accelerated corrosion. Injection of hydrogen peroxide has been explored as a possible mitigation for flow accelerated corrosion because most of the hydrogen peroxide partitions to the liquid phase and spontaneously decomposes into oxygen and water and thus enriches the liquid phase with oxygen. However, although the flow accelerated corrosion rate is decreased, hydrogen peroxide injection is not as effective as a remedy towards flow accelerated corrosion as replacement of materials to low alloy steel (alloyed with chromium) or the presence of a stainless steel coating.

— *Effect of alloy additions*. The flow accelerated corrosion rate is highest in carbon steel piping with very low levels of alloying elements. The presence of chromium, copper and molybdenum, even at low percentage levels, reduces the flow accelerated corrosion rate considerably. The relative corrosion rate of steels is reduced by 80% at a chromium content as low as 0.2%. The flow accelerated corrosion rate is decreased by a factor of 4 with the steel type 2–1/4 % Cr and 1% Mo (2–1/4 Cr-1 Mo steel). Austenitic stainless steels are virtually immune to flow accelerated corrosion.

— *The entrance effect*. In the 1990s a new flow accelerated corrosion wear effect was identified. It is referred to as the leading edge effect or the entrance effect. This effect occurs when flow passes from a flow accelerated corrosion resistant material to a non-resistant (susceptible) material, which causes a local increase in the corrosion rate. This effect is normally manifested by a groove up or downstream of the attachment weld between the corroding and the resistant material. In one relatively recent example, significant wear was detected in an expander.

Flow accelerated corrosion was considered to be a problem mainly in two-phase flow systems. A first case of single-phase flow accelerated corrosion induced pipe failure was reported in 1985 when the Trojan NPP experienced catastrophic failure of a DN350 heater drain pump discharge pipe made of SA-106 Grade B carbon steel. The failure caused the release of a steam-water mixture of approximately 180°C into the turbine building. In addition to the fire suppression system actuation by heat sensors in the turbine building and damaged secondary plant equipment, one member of the operating staff received first and second degree burns on 50% of his body from the high temperature fluid. A second case of single-phase flow accelerated corrosion induced pipe failure was reported in 1986 when a DN450 suction line to the main feedwater pump at Surry-2 failed in a catastrophic manner. The line temperature at this location was approximately 185°C, with a pressure of approximately 2.6 MPa.

These two events are of historical significance. They demonstrated that significant flow accelerated corrosion induced pipe wall thinning can occur not only in wet steam lines (two-phase flow conditions) but also under single-phase flow conditions. From a flow accelerated corrosion management perspective, the two events raised questions about the effectiveness of the then existing (mid-1980s) non-destructive examination programmes to monitor piping integrity for wall thinning and prevention of pipe failure.

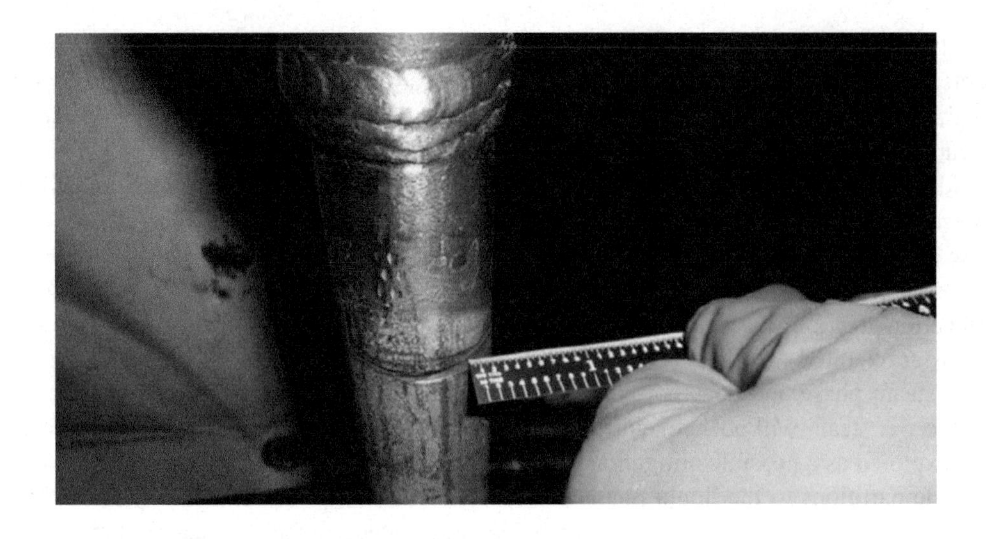

FIG. 12. Pressurizer (PZR) spray line wear caused by flow induced vibration (reproduced from [47]).

2.9. FLOW INDUCED VIBRATION AND FRETTING WEAR

Flow induced vibration may lead to fatigue and loss of material due to fretting (or abrasive wear) [45, 46]. Flow induced vibration is the most important degradation mechanism affecting steam generator tubes and heat exchanger tubes. Piping systems are also affected by flow induced vibration, especially in locations where there are inadequate supports or insufficient clearance (or gaps) between piping and adjacent components or structures. Examples of such material loss include: (a) interaction between reflective metal insulation end caps and piping (Fig. 12), and (b) interaction between grating steel bars and piping. The latter form of interaction has resulted in significant pipe failures.

2.10. STRESS CORROSION CRACKING

Stress corrosion cracking or environmentally assisted cracking is mainly observed in the weld deposit and heat affected zone and it is considered that it occurs due to the synergistic effect of three factors of material, stress and environments. SCC may occur when a susceptible material is subjected to stress in a corrosive environment. One example of a scenario that might lead to SCC is one in which a weldment is sensitized due to high heat input, subjected to high local stresses such as welding residual stresses, and the weldment is subjected to a corrosive environment. Without mitigation, there are four types of SCC mechanisms that are acting on WCR piping:

— Intergranular SCC of stainless steel;
— Primary water SCC, intergranular SCC that occurs during exposure of nickel base alloys to high temperature PWR primary water;
— Transgranular SCC, including external chloride induced SCC of stainless steel;
— Strain induced corrosion cracking of high strength carbon steel.

2.10.1. Intergranular stress corrosion cracking

The intergranular SCC of stainless steels is a time dependent type of material degradation phenomenon. Its morphology is associated with the temperature/time fabrication conditions that gave rise to thermal sensitization and the formation of chromium carbide precipitation (e.g. $M_{23}C_6$) and chromium

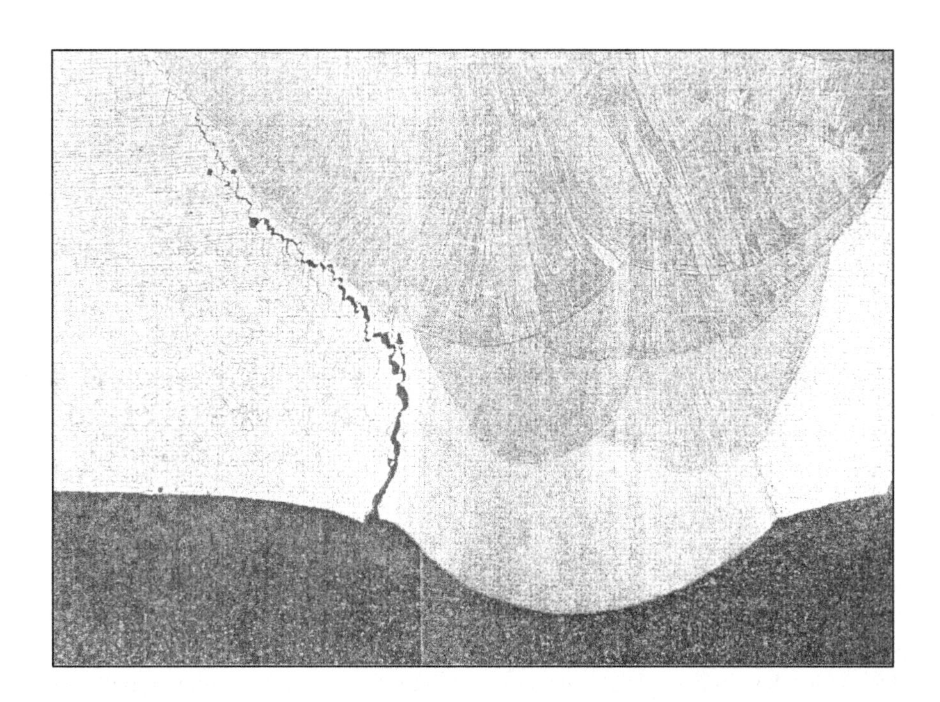

FIG. 13. Intergranular SCC in a weld heat affected zone (reproduced from [48]).

depletion at the grain boundary. The reduction in chromium concentration adjacent to the grain boundary gives rise to a reduction in passivity and makes the material susceptible to intergranular SCC. The cracks originate from the water side of the pressure boundary. It is the BWR aqueous environment which is responsible for the corrosion factor in the cracking as illustrated in Fig. 13.

Subsequent to the introduction of low carbon and stabilized grades of stainless steel, intergranular SCC has occurred in these materials that were clearly not in a sensitized condition. It has been shown that their susceptibility to intergranular SCC is due to cold work induced during fabrication. In many cases, the initial cracking was found to be initially transgranular then changing to an intergranular cracking mode. The initial transgranular cracking is often associated with a surface layer of cold work induced by grinding.

2.10.2. Primary water stress corrosion cracking

The intergranular SCC of Ni base alloys is a time dependent type of material degradation phenomenon. Nickel base alloys, particularly Alloy 600, and weld metals 82, 132 and 182, have proved to be generically susceptible to intergranular SCC in normal specification PWR primary water systems. This is commonly referred to as primary water SCC. The operational experience shows that the fabrication induced residual stresses have a large influence on primary water SCC in Alloy 600 weld metal. Examples of components affected include RCS hot leg (Fig. 14), cold leg, drain and reactor coolant pump nozzle to safe-end dissimilar metal welds.

Primary water SCC in the weld metal grows along the grain boundaries of columnar crystal dendrite packets. Initiation in the weld metal is often thought to be the result of typical and non-typical fabrication processes leading to locally high residual stresses, or surface stresses from, for example, grinding. To date, it has been found that the susceptibility of nickel base alloy weld metal to SCC is higher than that of the base metal. Intergranular SCC of nickel base alloys in BWRs is believed to be attributed to chromium depletion at grain boundaries, similar to intergranular SCC in thermally sensitized stainless steels.

Alloy 600 (or Inconel 600), a nickel base metal, was developed in the 1950s for use as a construction material for NPPs. The material was qualified for use in NPPs because of its perceived resistance to SCC; it was viewed as an alternative to Type 304 or Type 316 austenitic stainless steels. An early (possibly

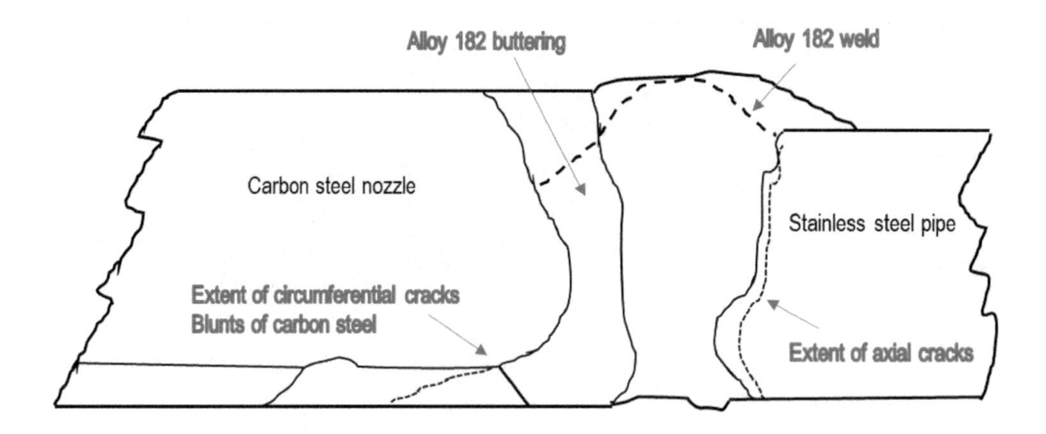

FIG. 14. V.C. Summer NPP RCS hot leg to reactor pressure vessel dissimilar metal weld cracking.

the earliest) recorded instance of SCC of Alloy 600 material is that of the failed inspection tubes in the Swedish Ågesta Reactor[3] in September 1964.

Alloy 690, which has higher chromium content than Alloy 600, was developed in 1960s and has been widely used since the 1980s to replace Alloy 600 components in PWRs, beginning with thin walled steam generator tubes, and eventually including thick section nozzle penetrations in reactor pressure vessel heads. This choice was made following numerous laboratory studies which confirmed the resistance to SCC in reactor coolant primary water for this alloy [49]. Alloys 52 and 152 are the higher Cr content nickel base weld metals that are used for joining Alloy 690 components, and which are also used for weld overlays to mitigate the SCC susceptibility of Alloy 82 and 182 welds. To date, there are no known cases of in-service primary water SCC of Alloys 690, 52 and 152.

2.10.3. Transgranular stress corrosion cracking

The earliest indications of cracking in unirradiated austenitic stainless steels occurred in the late 1960s in components where the temperature was <100°C and this was observed during storage and fabrication, and operation. The degradation mode was transgranular SCC on the outside surface of the pipe, as shown in Fig. 15. Transgranular SCC of austenitic stainless steels is mainly caused by chloride contamination, although other halide anions, such as fluorides, can induce it as well. The problem initiates on the outside surfaces of austenitic stainless steel components, mainly owing to a lack of adequate cleanliness. Wetting due to condensation or nearby water leaks can form an aqueous environment leading to transgranular SCC accompanied by pitting or crevice corrosion. Implementation of known procedures that ensure adequate surface cleanliness at all stages of construction and operation of NPPs is a continuing necessity and requires adequate management.

Chloride induced transgranular SCC can take place in internal surfaces, generally in dead legs and stagnant regions due to the high probability of the simultaneous presence of chloride contamination and oxygen. The canopy seals, which ensure the pressure boundary of threaded connections in the PWR control rod drive housings located in the reactor pressure vessel head, are the areas that have been rather frequently affected by transgranular SCC. Leaks from the canopy seals have caused serious boric acid corrosion of the upper head low alloy steel.

Transgranular SCC has also occurred from inner surfaces, mainly in pipe sections containing stagnant two-phase coolant, where evaporation and concentration of chlorides can occur. Wetting due to condensation or nearby water leaks creates an aqueous environment leading to transgranular SCC, usually

[3] A combined district heating and power reactor sited below ground near Stockholm, Sweden. The reactor was permanently shut down in 1974.

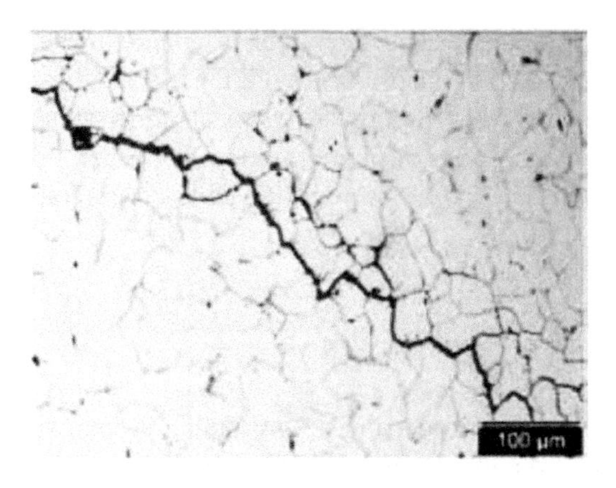

FIG. 15. Transgranular SCC in the base metal of a pipe section (adopted from [50]).

accompanied by pitting or crevice corrosion. The stress required for chloride induced transgranular SCC is relatively modest with its threshold close to the proportional yield strength of solution-annealed austenitic stainless steels. Implementation of the known adequate procedures to ensure appropriate surface cleanliness at all stages of construction and operation of NPPs is a continuing necessity that requires careful management attention.

External chloride induced SCC is transgranular SCC initiated on the outside surface of a component due to the presence of chloride in sea salt, coatings, etc., attached to the material surfaces and by perspiration. For external chloride induced SCC to take place several factors are usually present: high carbon content, tensile stress, moisture, heat, an aggressive chemical environment (e.g. chlorides or fluorides) and sensitized metal. Heat treatment or welding sensitizes the metal by depleting the chromium in the grains and forming chromium compounds at the grain boundaries.

2.10.4. Strain induced corrosion cracking

Strain induced corrosion cracking is used to refer to those corrosion situations in which the presence of localized dynamic straining is essential for crack formation to occur, but in which cyclic loading is either absent or restricted to a very low number of infrequent events. Strain induced corrosion cracking has been observed in pressurized components in German NPPs made of higher strength ferritic carbon steel. This kind of degradation has caused circumferential cracking in feedwater nozzle regions and at welds and axial cracking in pipe bends but also in straight sections of thin walled piping in German BWRs. The use of fine grained steels such as WB35 and WB36 allowed the use of thinner walled piping without stress relief treatment of welds [51]. The features that aggravated the cracking susceptibility in these incidents, based on [52, 53], are as follows:

- Dynamic straining associated with, for instance, reactor startup or thermal stratification during low feedwater flow or hot standby conditions are leading to a wide range of applied strain rates, which are expected to increase the crack propagation rate.
- High local stress at or above the high temperature yield stress in so influencing a lack of plastic constraint at the incipient crack tip; accordingly, an anomalous increase in crack propagation rate occurs due to an effective increase in crack tip strain rate. In the failure analyses, such high local stresses were attributed to weld defects (e.g. misalignment of weld edges), piping fit-up stresses and, in some cases inadequate pipe support at elbows. The combination of this high stress adjacent to the weld and the high applied strain rate led to a distribution of multiple cracks around the circumference of the pipe that was no longer confined by the asymmetric azimuthal distribution of weld residual stresses.

— Oxidizing conditions, in combination with intermediate temperatures and potential anionic impurities, may influence coolant conductivity and its pH within a crack. This combination of environmental factors was further intensified when, during the reactor shutdown, stagnant water was sometimes left exposed to air in horizontal portions of piping; pitting and general corrosion occurred under these low temperature conditions, where pits were observed to act as crack initiators during subsequent operating cycles.

2.11. HYDRAULIC PRESSURE TRANSIENTS

Water hammer is a phenomenon occurring in any piping system with valves used to control the fluid (or steam) flow; it occurs from the starting and stopping of pumps, from the opening and closing of valves, from changes in flow direction or from water column separation and collapse [54–60]. It is a result of a pressure surge (high shockwave) that propagates through a piping system because fluid in motion is forced to change its direction or abruptly is forced to stop moving. This abrupt change in fluid momentum makes a shock wave travel back and forth between the cause that created it and a resistance point in the system such as a restriction orifice, a valve downstream or a number of elbows in between. If the intensity of a shock wave is high, physical damage to the system can be expected. Ideally, a piping system is designed to absorb pressure transients or to attenuate them. Although the water hammer experience is extensive, relatively few water hammer events have resulted in catastrophic pipe failures.

2.12. HYDROGEN COMBUSTION IN PIPING

Hydrogen is used in WCR plants for primary water chemistry control and as coolant for electric generators. It is stored as high pressure gas in vessels and is supplied to the various systems in the auxiliary building, reactor building and turbine building through small diameter piping. Leaks or breaks in the piping can result in the accumulation of a combustible or explosive mixture of air and hydrogen within a building structure. A hydrogen-oxygen gas mixture is called radiolysis gas if it is generated by dissociation of water under the influence of gamma and neutron radiation. Radiolysis gas can appear, for example, in safety relevant piping of NPPs. A summary of observed hydrogen ignition mechanisms is well described in [61].

In BWR plants the radiolysis gas is entrained in the main steam flowing to the condenser; there, it is exhausted and recombined to become water again. However, the radiolysis gas entrained in the steam flow also reaches into plant components that are connected to the primary system. Under certain favourable conditions, the radiolysis gas can accumulate in areas of stagnant flow strongly favoured by steam condensation or by evaporation due to a pressure drop. As the accumulated radiolysis gas is present at an ideal stoichiometric ratio, an accumulation of radiolysis gas may result in detonation. This presupposes that there is an ignition mechanism such as a pressure surge or heat-up upon valve operation.

Radiolysis gas also forms in the PWR core. Hydrogen is added to its primary coolant to optimize the water chemistry which increases bonding of oxygen resulting in practically no radiolysis gas present at an ideal stoichiometric ratio. Hydrogen is then released again together with other gases into coolant storage tanks. In order to avoid formation of explosive hydrogen/oxygen mixtures in these vessels, the vessel atmosphere is continuously purged with nitrogen. A unique aspect of water cooled, water moderated power reactor (WWER) plants is the use of ammonia in the primary water, which decomposes in the reactor radiation field to form hydrogen and nitrogen.

2.13. HYDROGEN EMBRITTLEMENT

Hydrogen embrittlement, also known as hydrogen assisted cracking, involves the ingress of hydrogen into a component, an event that can seriously reduce the ductility and load bearing capacity, cause cracking and catastrophic brittle failures at stresses below the yield stress of susceptible materials. Hydrogen embrittlement occurs in a number of forms, but the common features are an applied tensile stress and hydrogen dissolved in the metal. This form of material degradation has resulted in a 'break before leak' type of pipe failure. Metallurgical analyses of failed components have found transgranular cleavage on the fracture surface, high hardness values in the region exposed to the process fluid and a hydrogen rich environment, which are all consistent with hydrogen embrittlement.

2.14. THERMAL AGEING EMBRITTLEMENT

Cast austenitic stainless steels are used to produce pipe elbows for use in RCSs of WCRs. The cast components can suffer a loss in fracture toughness due to thermal ageing embrittlement after many years of service at temperatures in the range of 280–320°C. Thermal ageing of cast stainless steels at these temperatures causes an increase in hardness and tensile strength and a decrease in ductility, impact strength and fracture toughness of the material. There are no known failures (as in through-wall leaks) that are attributed to thermal ageing embrittlement, however.

The magnitude of the reduction of fracture toughness depends on the type of casting method, the material chemistry and the duration of exposure at operating temperatures conducive to the embrittlement process [62–64]. Static castings are known to be more susceptible than centrifugal castings, high molybdenum content castings are more susceptible than low molybdenum content castings, high delta ferrite castings are more susceptible than low delta ferrite castings, and higher operating temperatures increase the embrittlement rate compared to the rate at lower operating temperatures (285°C). The extensive amount of fracture toughness data available for thermally aged cast austenitic stainless steels materials enables delta ferrite, molybdenum content, casting type and service temperature history to be used as the bases for screening and evaluating components for operation beyond 40 years.

3. PIPING RELIABILITY ANALYSIS FRAMEWORK

3.1. OVERVIEW

This section presents a framework on how to organize a piping reliability analysis task. The framework has evolved over a period of multiple decades and it has benefited from technical insights gained from practical applications performed in different contexts and by different analysts. The framework consists of seven steps (or activities) as shown in Fig. 16. An objective of the analysis framework is to promote consistency in how an analysis is organized, executed and documented. The seven steps are as follows:

(1) *Define the evaluation boundary.* An evaluation boundary defines the piping system and its components (e.g. bends or elbows, pipes, tees, welds) for which reliability parameters are to be derived. This step is explained in Section 4.

(2) *Identify the potential and observed degradation mechanisms.* Degradation mechanism assessment is a formal, systematic process for identifying the possible degradation mechanisms and it is based on many decades of accumulated knowledge about the conjoint requirements for pipe degradations. This step is explained in more detail in Section 5.

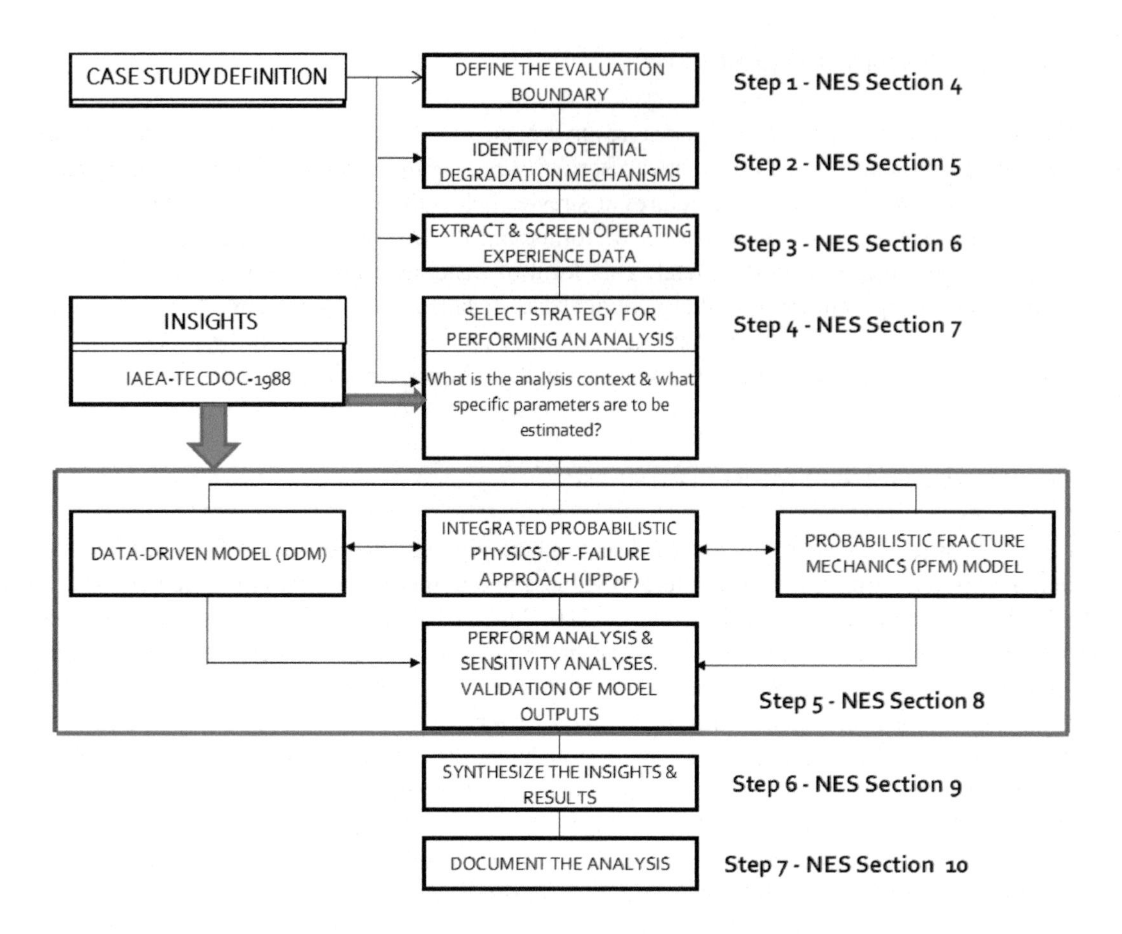

FIG. 16. Framework for performing piping reliability analysis.

(3) *Extract and screen the OPEX data* that correspond to the evaluation boundary. Data driven models (DDMs) of piping reliability rely on real-life observations of pipe degradation and failure over time. The pipe failure OPEX data to be extracted from a pipe failure database correspond to the evaluation boundary for which the reliability parameters are to be determined. However, regardless of the selected quantification scheme, the analyst is expected to examine the existing, relevant advanced WCR/WCR OPEX to determine if there is a basis for reconciling model output with real data. Annex III provides supporting information with reference to possible sources of pipe failure and piping reliability data. Step 3 of the analysis framework is explained in more detail in Section 6.

(4) *Select the strategy* for performing an analysis. Based on a specific evaluation boundary, in this step the calculation case(s) and the specific reliability metrics are defined. Supporting engineering calculations are identified for the purpose of calculation case definition such as specific pipe failure modes to be considered and with reference to potential spatial impacts as characterized by a zone of influence. Examples of engineering calculations include transient thermal-hydraulic analysis and simulation. Leak rate calculations are done on the basis of equivalent break size (EBS), crack opening displacement or crack opening area [62, 63]. Section 7 includes a discussion on methodology selection criteria that are based on insights from practical applications as well as insights obtained from benchmark studies. The inter- and intra-comparisons of the benchmark results as described in [24] provide information on the strengths and limitations of different methodologies when applied to a common evaluation boundary.

(5) *Perform the analysis* by applying one or more methods and methods implementations (i.e. different data driven methodology, probabilistic fracture mechanics (PFM) methodology and a physics based

approach, integrated probabilistic physics-of-failure (I-PPoF)). This step is explained in further detail in Section 7:
  — Define input parameters;
  — Document assumptions;
  — Determine the types of sensitivity analyses to performed;
  — Perform an integrated uncertainty analysis and validate the model output.

(6) *Synthesize the insights and results*. Certain follow up (or sensitivity) studies may have to be performed once a base case set of probabilistic failure metrics has been obtained. Sections 8 and 9 provide relevant information with respect to results validation and interpretation.

(7) *Document the analysis*. The documentation consists of assumptions, input parameters, model outputs and results interpretation. End user requirements on an analysis affect how the results are processed and documented. This step is explained in further detail in Section 10.

3.2. NOMENCLATURE

A central aspect of the analysis framework is the importance of using a consistent terminology. Especially with respect to (a) the definition of what constitutes a pipe failure, (b) the minimum requirements to be placed on a state of the art piping reliability model, (c) differentiation between pipe failure rate, pipe failure frequency and pipe failure probability, and (d) the different structural integrity management processes and how these could be addressed analytically.

3.2.1. Pipe failure mode definitions

In applying and comparing the results obtained from different piping reliability models, the definition of what constitutes a failure becomes important. It is noted that ambiguity exists in the use of pipe failure mode terminology. Fundamentally, the term 'failure' implies that the integrity of a pressure boundary is compromised. The manner in which pipe material degrades and the consequence of a degraded condition influences how structural reliability is modelled.

The term 'pipe failure' refers to a degraded pressure boundary. It can be a rejectable flaw or structural failure. The rejectable flaw does not meet the requirements of applicable national codes and standards. The structural failure directly impacts the NPP operation (e.g. safety system actuation with reactor trip), with potentially significant dynamic impacts on adjacent NPP structures, systems or components. The input parameters to analysis are the flaw location within a piping system, its size and orientation. When classifying OPEX data, the following terminology is used:

- Rejectable defect or flaw requiring repair or replacement:
  - Weld repair to provide a leak barrier;
  - Code repair that involves radiography or ultrasonic examination to verify the integrity of a weld repair;
  - Full structural weld overlay as means for arresting future crack growth;
  - Replacement in-kind using the same material and configuration;
  - Replacement using new material;
  - Replacement including re-routeing using the same material to alleviate pipe stress risers and minimize fatigue vulnerability;
  - Replacement including re-routeing using new material to eliminate or minimize pipe stress risers and minimize fatigue vulnerability.
- Through-wall flaw, inactive leakage; detected during visual inspection.
- Through-wall flaw, active leakage; detected by leak monitoring system or during periodic walk-down inspection.

The international and national codes and standards include details on pipe failure mode determination and the methodology to be applied when assessing the fitness for continued operation given a degraded piping pressure boundary. Crack growth is characterized as either subcritical/stable crack growth or critical/unstable crack growth. A subcritical crack is a crack whose stress intensity factor is below the critical value. Typically, it is assumed that fatigue cracks and stress corrosion cracks are related to the stress intensity factor, *K*, unless under extreme loading scenarios such as water hammer or beyond design basis accidents where large plastic deformation occurs. The stress intensity factor is used in fracture mechanics to predict the stress intensity near the tip of a crack caused by a remote load or residual stresses. If the energy available for an incremental extension of a rapidly propagating crack falls below the material resistance, the crack arrests.

In PFM, a leak is a wall-penetrating defect which is stable. A wall-penetrating defect which fails due to the applied loads is considered as a rupture with unstable crack growth rate.

In PSA, the definition of the consequence of a pipe failure is very important; however, questions such as 'what is the frequency of a pipe leak?' or 'what is the frequency of pipe rupture?' are not meaningful. In probabilistic terms, the definition of a failure versus its consequences affects the modelling approach as well as the modelling of uncertainty. The consequence of a pipe failure can be characterized in terms of through-wall mass or volumetric flow rate, kg/s and m^3/s, respectively. The size of a through-wall pipe flaw is also used and expressed in terms of an EBS or crack opening area. The EBS is calculated based on the engineering analyses that rely on fluid dynamics; in other words, considerations of the consequence of a potential pipe failure that determines how it is modelled and how the uncertainties are characterized, with an example shown in Fig. 17 developed based on information provided in [24].

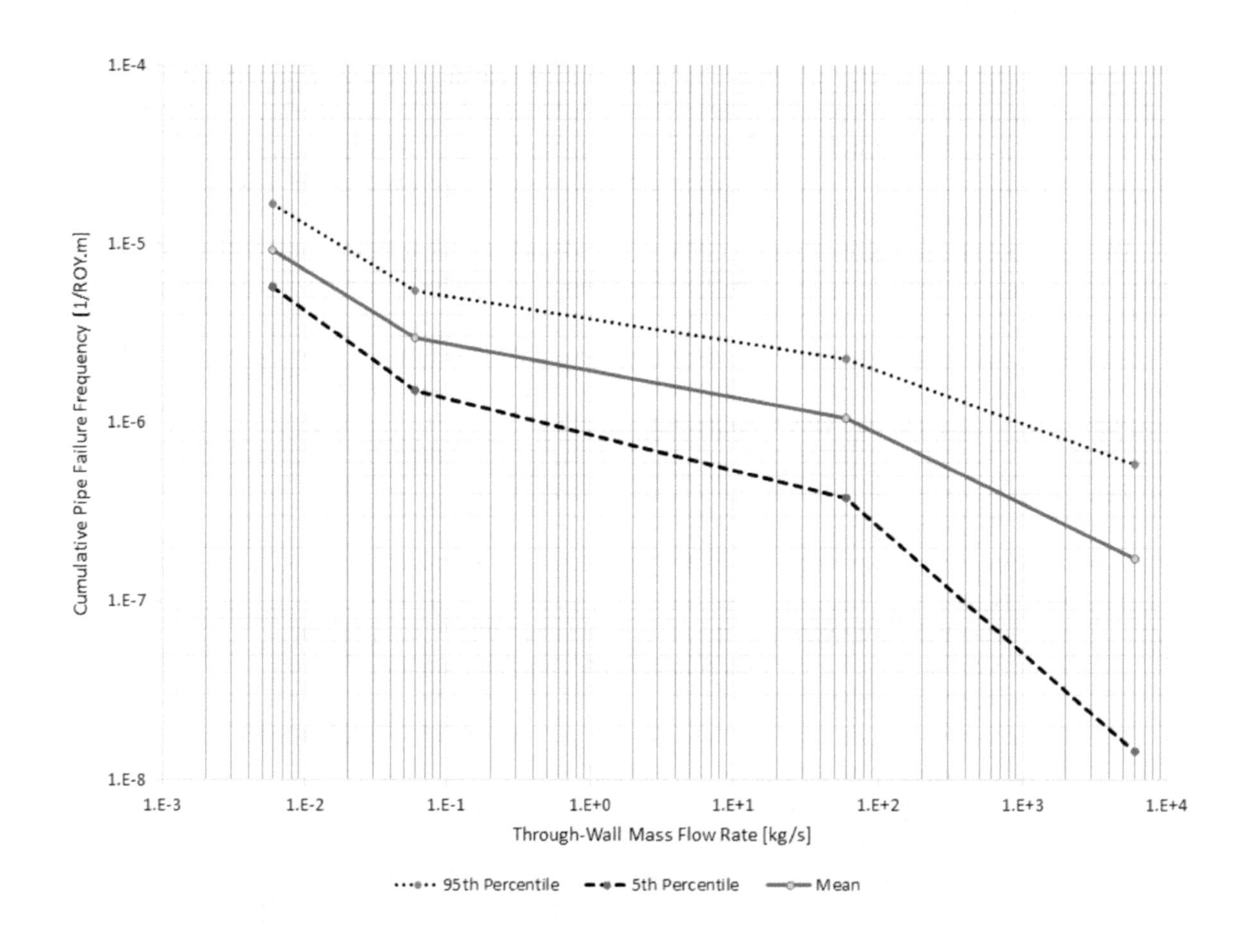

FIG. 17. Frequency of pipe failure as a function of through-wall flow rate.

Pipe failure frequencies are calculated for different initial flaw sizes, consequential through-wall flow rates and accounting for the effectiveness of ISI through different assumptions about the probability of detecting a pre-existing flaw. As indicated in Fig. 17, the OPEX data (i.e. number of observed failures) that underlie the smaller consequences (<10 kg/s through-wall flow rate) tend to be more robust than for the major, energetic structural failure event that produces through-wall flow rates greater than 100 kg/s.

The crack opening area is a key element in fracture mechanics assessments [65, 66]. Estimates of the crack opening area for a postulated through-wall cracking can vary widely depending on how the crack is idealized, which crack opening model is used and what material properties are assumed. A wide range of published solutions is available for idealized notch-like cracks in simple geometry subject to basic loading (pressure, membrane and bending). Their accuracy varies with geometry (e.g. pipe diameter and pipe wall thickness), crack size, type of load and magnitude of load.

3.2.2. Treatment of uncertainties

The integrated consideration of uncertainty is a necessary element of any piping reliability analysis, irrespective of the chosen methodology. The event sequence diagram in Fig. 18 is a simplified representation of the piping integrity risk triplet. It addresses what can go wrong, how likely it is to happen, and what the consequences are of a structural failure. Each element in the risk triplet is represented by a probability density function [67], or multiple probability density functions, depending on the desired or required level of granularity of an analysis.

Piping reliability analysis estimates the frequency of failure and conditional probability of failure. Depending on the objective of an analysis, the term 'failure' has different definitions, oftentimes colloquially referred to as breach, break, leakage or rupture. As previously indicated, in the context of structural integrity analysis and PSA applications any of these terms are meaningless unless further qualifications are provided. Of essence is the estimation of the frequency of a degraded structural integrity state as a function of its consequence in terms of the observed mass or volumetric through-wall flow

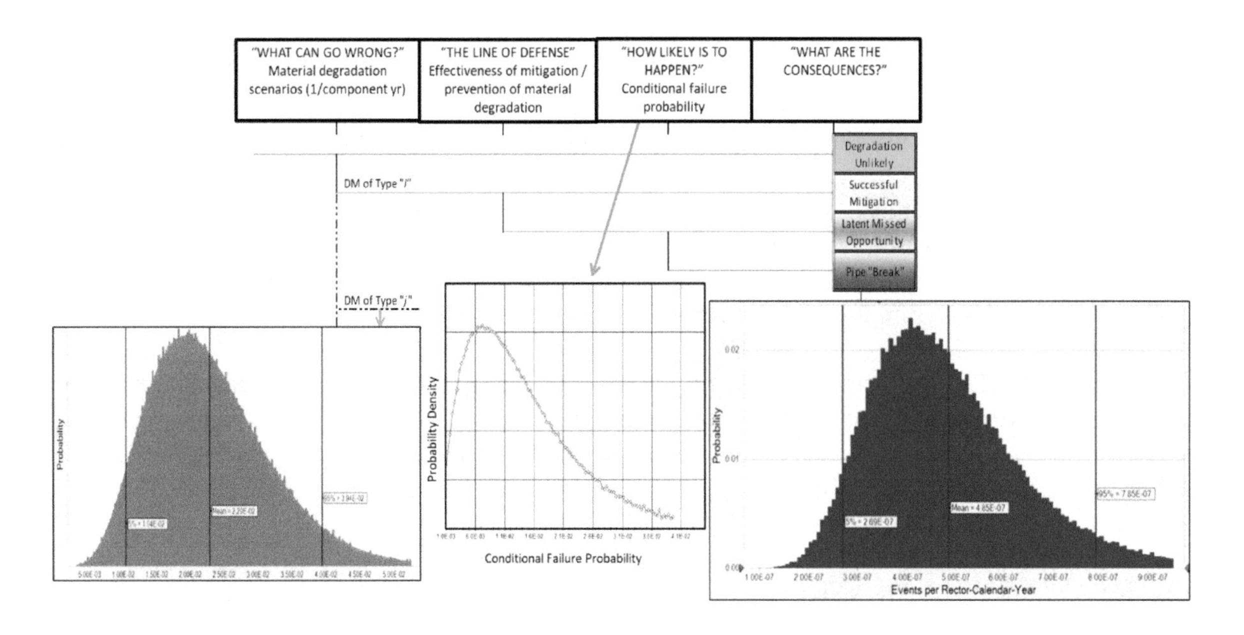

FIG. 18. Structural integrity risk triplet.

rates. From codes and standards and regulatory perspectives, here are some examples of pipe failure consequences and their effect on decision making:

— A through-wall leak originates from a crack or material loss that goes all the way through the pipe wall. If it is in safety related pipe, the through-wall leak needs to be repaired and within specific time limits and according to codes and standards.
— If the through-wall leak is in the primary coolant system and the coolant loss exceeds the detection limit of about 5×10^{-3} kg/s, a stepwise approach with action levels is to be implemented by the plant. The plant specific technical specifications include details on the leak detection limits.
— If the through-wall leak is in a moderate energy piping system, the applicable codes and standards may include options for applying a temporary repair.

A common practice in presenting the analysis results is frequency (y axis) versus consequence (x axis) (e.g. through-wall leak rate or EBS). A higher frequency is normally associated with smaller size piping or smaller EBS, and less uncertainty. There is not a universally accepted presentation format, however. Numerous studies have been performed to investigate the relationship(s) between pipe diameter and the calculated frequency of pipe failure. Advanced PFM methods, probabilistic physics-of-failure methods and/or statistical models could be applied to obtain insights into this matter.

3.2.3. Structural integrity management

The structural integrity of piping components and systems is monitored and managed through mandated and voluntary RIM programmes. The mandated programmes reflect good engineering practices, OPEX insights, and material science and structural reliability advancements, and national codes and standards. The voluntary or owner defined RIM programmes augment the mandated (or codified) programmes to account for plant specific operating conditions and unique piping design features. Examples of RIM activities are:

— Scheduled, periodic visual inspections and walk-down inspections;
— Means for detecting and locating pressure boundary leakage (e.g. flow monitoring, mass balance calculation, heat sensors, visual inspection);
— Fatigue monitoring (e.g. vibratory fatigue and thermal fatigue loads);
— Monitoring of material degradation such as pipe wall thinning caused by flow accelerated corrosion, erosion-cavitation, erosion-corrosion, microbiologically influenced corrosion;
— Non-destructive examination using different technologies including eddy current, radiography and ultrasound;
— Non-destructive examination qualification; personnel and technology;
— Water chemistry control; including primary and secondary side water, cooling water;
— Pipe stress improvement using different technologies including full structural weld overlays, induction heat stress improvement, mechanical stress improvement process.

The effects of RIM on structural integrity are always accounted for in piping reliability analysis; either explicitly or implicitly. Some RIM processes are implemented to detect flaws in piping before an active degradation mechanism produces a leak. Other RIM processes are implemented to mitigate or prevent future material degradation.

3.3. MEANING OF LEAK AND FAILURE

The investigation of pipe failure rates involves assessing the occurrence of failures; that is, significant and measurable losses of structural integrity associated with defects, and with adverse effects to the safety and reliability of an NPP. The perceived and intended meanings of pipe failure definitions are addressed in more detail in this report section. Most prominent failures in piping systems are differentiated as leaks or ruptures. Oftentimes the exact meaning of these terms is not discussed further. However, these terms appear in different contexts and are therefore not interchangeable, which motivates a further examination of these commonly used terms. The four distinguishable fields in the following are: (1) the plant safety analysis, as done in PSA and thermal hydraulic simulations, (2) the structural mechanics analysis, (3) leak before break assessment, and (4) the actual plant operation together with the operational experience.

It is easier to start the discussion with rupture. A double-ended guillotine break has a very clear definition; it involves a complete separation of two pipe ends. This is a design assumption considered for the design capacity of the emerging core cooling system. A leak, in turn, is a reduced cross-section, from a thermal hydraulic perspective; it can be given as a fraction of the maximum, or directly as a cross-section. Related to this is the specification of a leak by the expected leak rate (at operating conditions), which allows one to quantify and compare the different leak sizes. This treatment is motivated by the fact that the reactor cooling system's behaviour in the case of a coolant loss rate is the object of interest.

From the structural mechanics analysis perspective, failure is associated with a spontaneous structural degradation when the material strength cannot withstand the loads and stresses. Three types of failure are relevant for piping:

— Failure of an intact pipe;
— Failure of a pipe with a surface flaw;
— Failure of a pipe with a through-wall defect.

Spontaneous failures of intact pipes are known to occur from overpressure or water hammer and take the form of extended axial bursts, but the mechanical aspects are entirely covered by design rules. More emphasis is on the analysis of defects in pipes for failure analysis, since highly reliable components show a safe behaviour even in the presence of defects. Within all possible defects present in metallic piping, crack-like defects are of special interest, since sharp tips of cracks lead to stress concentrations, and failure corresponds to an uncontrolled rapid growth of a crack if the stress concentration exceeds the material strength. For surface flaws, this means that the flaw grows until a leak is formed, and the result is then a crack-like wall penetrating defect. If this assessment is repeated with a wall-penetrating defect, the result would be an unstable growth of the flaw through the pipe, which is then associated with rupture. The practical end point of the unstable growth is usually not investigated, since this relies also on the stresses and their relaxation in the material: It is possible that a double-ended guillotine break is the result, but it would also qualify for a rupture if there remains a (small) connection between the pipe ends. The leak before break assessment is a procedure implemented in a number of standards as an additional safety demonstration [68, 69]. A central idea is that a small, local leak, which is not a threat to cooling the reactor, can be a precursor for a larger breach in the piping pressure boundary. Moreover, a small leak is potentially easier to detect than a surface flaw, since leak monitoring systems in a plant enable additional opportunities for the identification. Hence, a leak in the 'leak before break' context is assumed to have a leak rate in the order of the plant's detection threshold, which is often assumed to be around 0.061 kg/s (or about 1 gpm). The rupture instead is something with a much larger leakage rate, where the stability of the local leak cannot be assumed any more. The adverse scenarios are that a local leak forms, its leak rate is too small to be detected until it becomes unstable, or if a surface flaw directly leads to a large rupture without an intermediate stable leak phase.

The natural definition of a leak is basically any loss of tightness, which might be due to damage of structures. Typical situations of leakage identification comprise increased sump rates, increased radioactivity, inventory loss or boric acid deposits in the case of PWRs. In contrast to the thermal-hydraulic approach and the leak before break assessment, the actual mass flow rate in actual leaks often can only be estimated roughly. A rupture, instead, would be associated with an event where the failure is immediately recognized, together with the consequences for further operation (see Fig. 19).

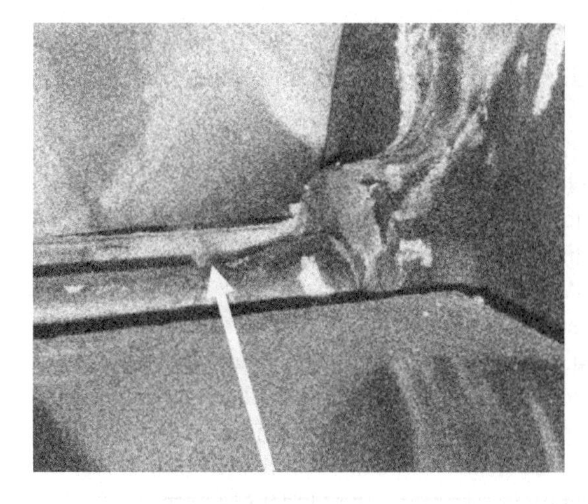

(a) Small leak

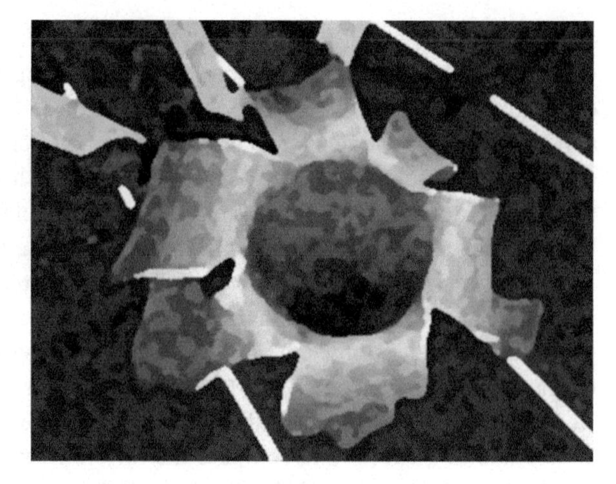

(b) Rupture

FIG. 19. (a) Small leak versus (b) rupture (adapted from Ref. [70] with permission).

In summary, the terms 'failure', 'leak' and 'rupture' are commonly used in different relevant backgrounds, with slightly different implications. While the thermal-hydraulic plant analysis and the leak before break assessment have the associated mass flow rate in the focus, the topological changes of flaws are central for the structural mechanics point of view. The definition by the event consequences is common to the operational standpoint and the probabilistic safety analysis. Although the definitions are not identical and the characteristics and differentiations vary, it is justifiable to use the simplified terminology within this report.

3.4. PRACTICAL ANALYSIS INSIGHTS

Piping reliability analysis can be computationally intense. As an example, in order to derive input to a PSA model, numerous calculation cases are defined to adequately cover the full range of different evaluation boundaries, degradation mechanisms and consequences of a pipe failure. Similarly, to address the uncertainty of the failure probability, a large amount of sensitivity studies and uncertainties analyses have to be performed. These are most computationally intense. Each calculation case is defined in terms of the desired model output. For example, pipe failure frequency versus consequence of a certain magnitude is usually characterized in terms of the size of a pressure boundary breach or the through-wall mass or volumetric flow rate. In the early PSA studies the different types of LOCA initiating events were given as small, medium and large LOCA; for example, 9.5 mm < small LOCA ≤ 35.5 mm EBS (i.e. diameter of circular hole) and 35.5 < medium LOCA ≤ 115 mm EBS. In support of a relatively recent WCR LOCA initiating event frequency estimation project, as many as 50 unique calculation cases had to be defined, for which up to 500 pipe failure reliability parameter distributions were generated to fully address the 'structural integrity risk triplet' [71].

4. EVALUATION BOUNDARY

4.1. DEFINITIONS

An evaluation boundary defines the piping system and its components (e.g. bends or elbows, pipes, tees, welds) for which reliability parameters are to be derived. The evaluation boundary definition is a first formal analysis activity of any piping reliability analysis task. The output from this task is necessary for specifying the analysis requirements (i.e. the input data and supporting engineering analyses) and the associated calculation cases.

4.2. INFORMATION SOURCES

Examples of required evaluation boundary information sources include process flow diagrams or piping and instrumentation diagrams[4], ASME Section XI (or equivalent) ISI programme plans, flow accelerated corrosion monitoring programme plans, general arrangement (or layout) drawings[5] and isometric drawings. There are three types of isometric drawings: (1) fabrication isometrics, (2) ISI isometrics, and (3) stress analysis isometrics. A fabrication isometric drawing consists of three sections: A main graphic section, which is a three dimensional representation of a pipeline route within a building structure, and it includes the following information:

- Pipeline number: usually an alfa-numeric designator that includes a system identifier and the nominal pipe size;
- Process flow direction;
- Pipe support locations;
- Piping component locations (e.g. bends, elbows, tees, branch connections);
- Weld locations; the weld identifiers include reference to the field weld and shop weld.

A title bar at the bottom of a drawing includes pipe line details such as line number, line size, insulation, operating and design pressure and temperatures, and pressure testing method. A section on the left or right side of the drawing consists of a bill of material (or parts list) section for the portion of a line shown in an isometric graphic. It includes the following information:

- Component description (e.g. long-radius elbow, 45° elbow, butt weld, etc);
- Component material and designation[6] (e.g. carbon steel type ASTM A-106 Gr. B, stainless steel 1.4436);
- Nominal size; pipe diameter and wall thickness;
- Welding process specification;
- Method of fabrication (e.g. welding technique, post-weld heat treatment);
- Number of pipe spools (or pipe sections).

---

[4] A piping and instrumentation diagram is a detailed diagram which shows the piping, tanks and vessels in the process flow, together with the instrumentation and control devices (e.g. check valves, flow control valves, isolation valves, pressure relief valves).

[5] An arrangement drawing shows the layout of a piping system and the connections to other SSCs such as pumps, tanks and valves.

[6] A corresponding material specification sheet provides the chemical and mechanical properties.

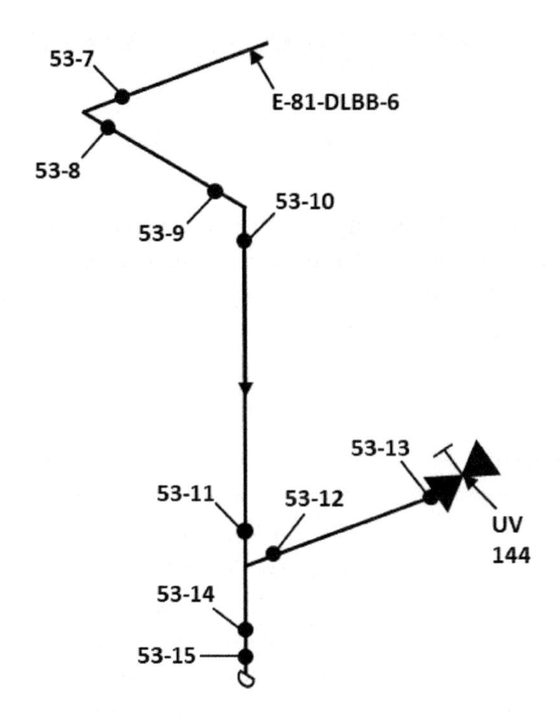

FIG. 20. Typical ISI isometric (adopted from widely published isomeric piping drawings).

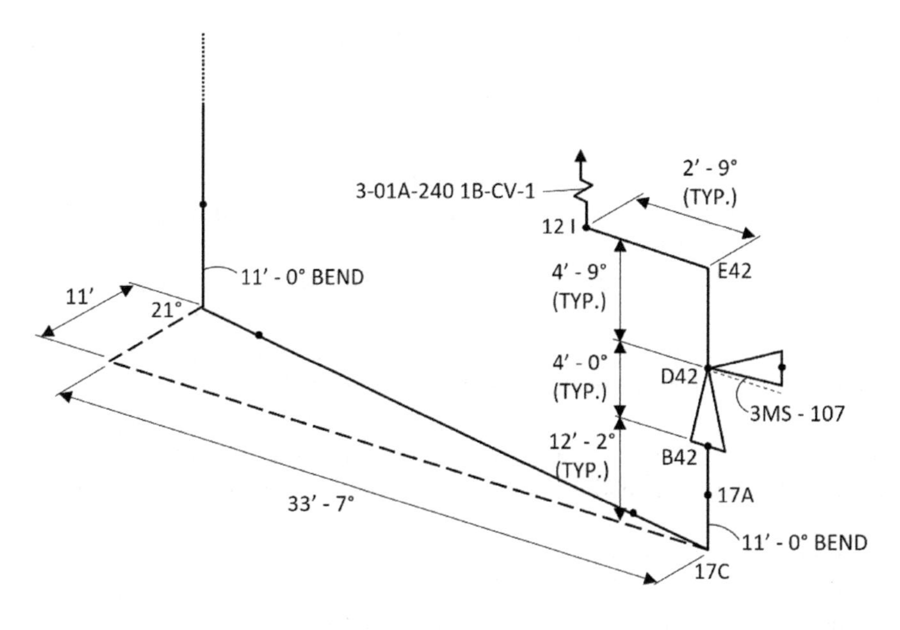

FIG. 21. Typical fabrication isometric (adopted from widely published isomeric piping drawings).

An ISI isometric drawing (Fig. 20) is a simplified version of a fabrication drawing. It includes line number(s) and the volumetric examination locations by weld identification numbers, weld details and pipe restraint locations as defined in an ISI programme plan.

The ISI isometric drawings sometimes are referred to as ISI programme plan zone drawings. A stress analysis isometric drawing is similar to a fabrication isometric drawing (Fig. 21), but it includes additional information relating to pipe support locations and how the supports respond to excessive or unusual piping displacements and loads. Included on a stress analysis isometric are the node numbers for which stress analyses have been performed. A piping layout drawing, or general arrangement drawing

(Fig. 22) shows all the major equipment, it is north/south and east/west orientation, and all piping leading to and from equipment are developed by piping designers. All of the main piping items (pumps, valves, fittings, etc.), instrumentation, access ladders and platforms are shown. The layout drawing usually shows a plan (top) view with elevations (side) and sectional drawings with piping dimensions and details including line numbers, size, specification, the direction of flow, etc. to help the piping designers extract all the necessary information for isometric or fabrication drawing preparation.

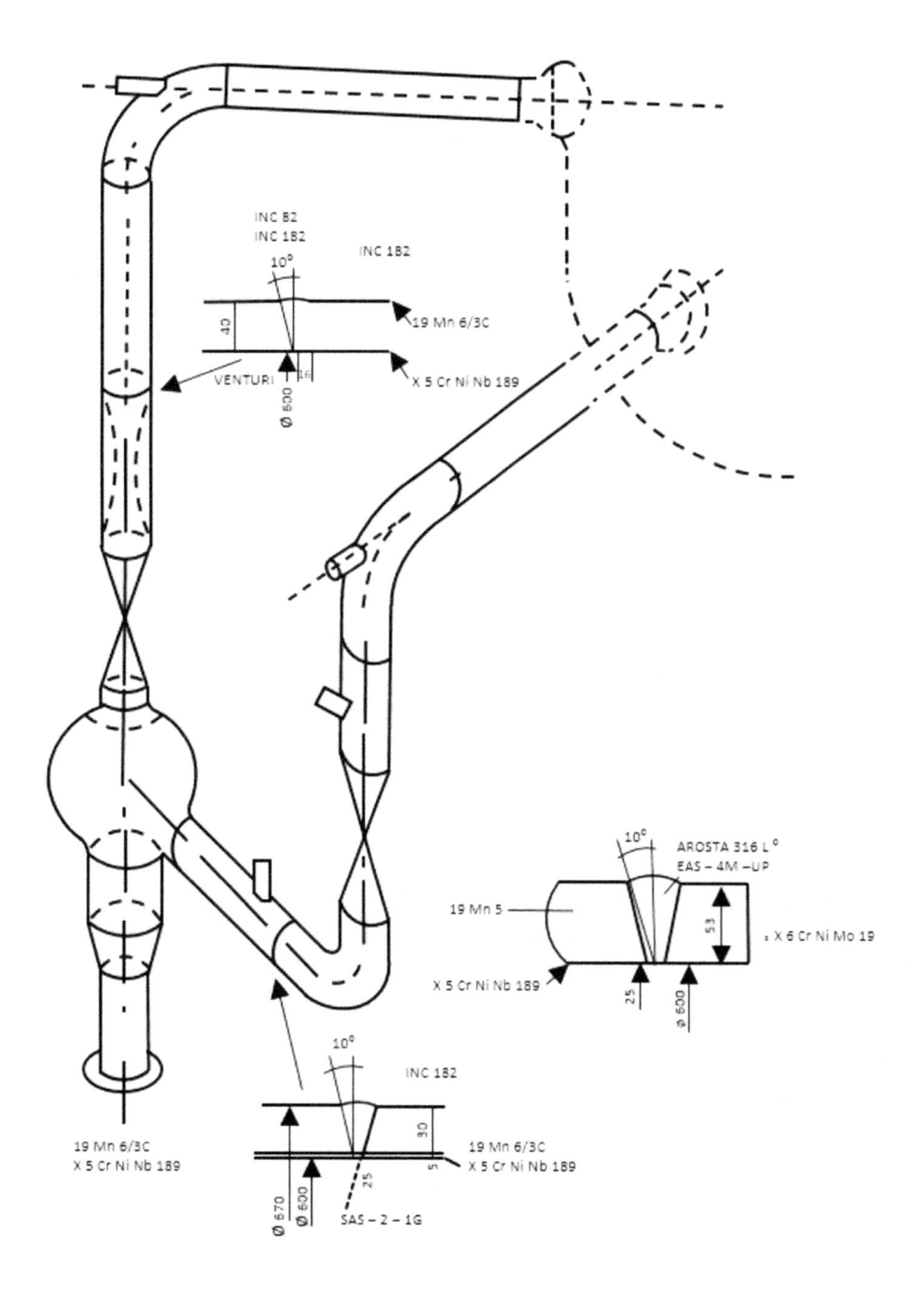

FIG. 22. An example of a piping layout drawing.

4.3. IN-SERVICE INSPECTION PROGRAMMES

The ISI programme plans document the inspection schedules and identify the components to be inspected, and how. An important part of the plans are the equipment drawings and isometric drawings. These drawings give the identification numbers and locations of welds, supports and other items that require periodic examination. Examples of other items include locations that have been known to be susceptible to a specific degradation mechanism such as thermal fatigue. A typical ISI programme plan includes safety class 1, 2 and 3 components. The information contained in these programme plans give the input to pipe failure rate exposure term determination, including a basis for assessing the plant-to-plant variability in weld populations.

4.4. FLOW ACCELERATED CORROSION PROGRAMME PLANS

Flow accelerated corrosion is known to affect carbon steel and low alloy steel piping components that operate in single-phase and two-phase flow conditions, with operating temperature over 100°C and low levels of dissolved oxygen. Predictive methodologies have been developed to determine flow accelerated corrosion wear rates (i.e. material loss) and to determine the remaining service life for each component included in an inspection programme. A typical flow accelerated corrosion programme plan includes a flow accelerated corrosion database, which gives details of the locations within a piping system that are susceptible to pipe wall loss caused by the chemical dissolution of the normally protective oxide layer formed on the surfaces of carbon and low alloy steel piping. Examples of piping system details in a flow accelerated corrosion database include:

- System name;
- Pipe line inside diameter and description (e.g. feedwater pump warm-up line, low pressure heater drain, high pressure heater vent);
- Nominal diameter and wall thickness;
- Material designation and chemical composition;
- Operating pressure and temperature;
- Component type (e.g. long-radius bend, 90° elbow, expander);
- Inspection data (e.g. date of first inspection, date of the most recent inspection);
- Operating hours;
- Flow at full power operation (e.g. continuous, intermittent);
- Type of fluid (e.g. steam, water/steam, water).

Querying a flow accelerated corrosion database yields summaries of the number of susceptible components organized by system, component type, diameter, wall thickness and material, etc. This information supports high energy line break analyses. A typical flow accelerated corrosion programme covers the following systems:

- Main steam including high pressure turbine exhaust;
- High pressure extraction steam;
- Low pressure extraction steam;
- Condensate;
- Feedwater;
- Moisture separator drains and vents;
- Feedwater heater drains and vents;
- Feedwater pump bypass;
- Steam dump to condenser.

The technical details of flow accelerated corrosion susceptible locations are documented in databases from which component population data can be obtained. For example, the number of 90° elbows in a given system and of a specific diameter, wall thickness and material composition.

4.5. ORGANIZATION OF PIPING SYSTEM DESIGN INFORMATION

The level of effort for defining an evaluation boundary is a function of the scope of an analysis. The scope may be limited to a uniquely defined single location within a specific piping system pressure boundary such as a high pressure safety injection branch connection on the cold leg of the RCS. When the scope of the analysis is to develop an initiating event frequency model for PSA, the evaluation boundary definition involves a detailed isometric drawing review to identify all possible pipe break locations. All pipe break locations are to be identified and evaluated for susceptibility to material degradation. Plants for which an approved risk-informed ISI programme plan has been implemented have additional relevant information sources that support the definition of an evaluation boundary. The risk-informed ISI programme documentation includes detailed weld lists organized by plant system, line inner diameters, diameter, weld type, material and operating conditions. Also included are the results of a degradation mechanism analysis performed for each inspection location. The degradation mechanism analysis considers the local operating conditions including chemical additives, flow conditions, pressure and temperature, and potential for thermal stratification and water hammer.

4.6. ADVANCED PIPING DESIGN AND ANALYSIS

Advanced computer-aided design and computational fluid dynamics tools are used in the design of advanced WCRs. The former tool provides for interactive layout of piping in 3D computer models. The computer models allow for determining possible pipe to pipe interference, pipe to grating interference, personnel access for routine visual inspection and non-destructive examination, equipment removal spaces and construction access. Computer-aided design allows for automated generation of isometric drawings and bills of material for piping fabrication and installation. Many computer-aided design systems include interfaces with pipe stress analysis software systems and piping fabrication equipment such as numerically controlled pipe bending systems.

Computational fluid dynamics tools are used extensively in the visualization and analysis of dynamic effects of high energy pipe breaks and simulation of water hammer phenomena and resulting pipe stresses. It is a predictive tool as well as a tool for analysing how and why a pipe failure occurred [72, 73].

5. DEGRADATION MECHANISMS AND FAILURE MODES

5.1. MATERIAL DEGRADATION ASSESSMENT

The synergistic effects of off-normal operating and environmental conditions, and unusual or extreme loading conditions influence metallic piping degradation and failures. The triplet (material, environment, loading) represents the conjoint requirements for pipe degradation. Sometimes, subtle changes in any of the physical parameters embedded in this triplet, such as but not limited to pH, corrosion potential, H_2 content, temperature, flow rate, carbon content or post-weld heat treatment, can have a profound effect on the pipe degradation and its failure propensity. Therefore, piping reliability analysis is expected to reflect on basic understanding of the roles of, for example, metallurgy, water chemistry and pipe stresses in the achievement of high structural reliability. This principle is summarized in the scheme of conjoint factors defining material degradation, as shown in Fig. 23.

Degradation mechanisms of material can be classified in groups. While the aim of this chapter is not to the physics of material degradation, it is instructive to consider a classification scheme as outlined in Fig. 24.

Step 2 of the piping reliability analysis framework is concerned with the identification of degradation mechanisms that have the potential to act on specific evaluation boundaries. Degradation mechanism assessment is a systematic process for identifying the possible degradation mechanisms and it is based on many decades of accumulated knowledge about the conjoint requirements for pipe degradations. A process on how to perform degradation mechanism analysis is outlined in the non-mandatory Appendix

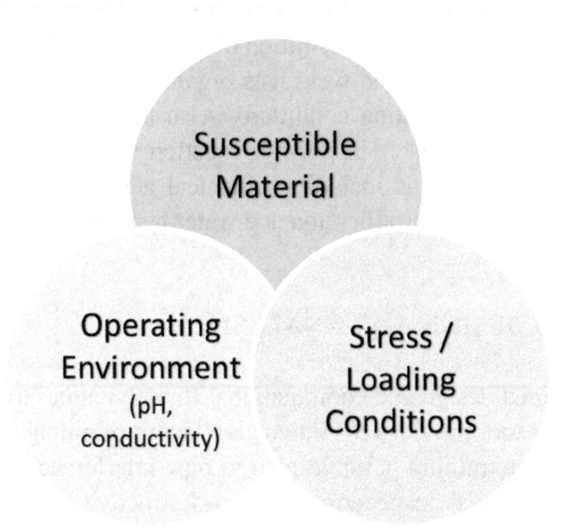

FIG. 23. Conjoint factors defining material degradation.

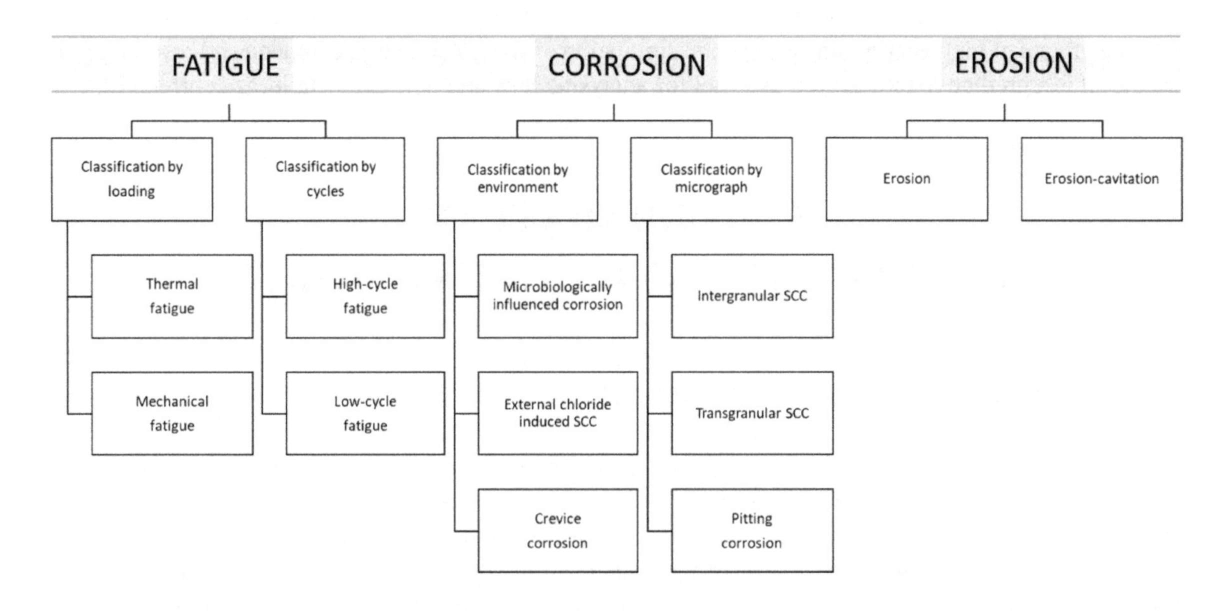

FIG. 24. Degradation mechanism classification scheme.

R of ASME Section XI [74]. In the context of RIM, a degradation mechanism analysis considers the following conditions:

— Design characteristics, including material, pipe size and wall thickness, component type and other attributes related to the system configuration (e.g. routeing, pipe supports).
— Fabrication practices, including welding and heat treatment.
— Operating conditions, including temperatures and pressures, fluid conditions (e.g. stagnant, laminar flow, turbulent flow), fluid quality (e.g. primary water, raw water, dry steam, chemical control) and service environment (e.g. humidity, radiation).
— Industry-wide service experience with the systems being evaluated.
— Results of pre-service, in-service and augmented examinations, and the presence of prior repairs in the system.
— Degradation mechanism analysis is performed by comparing design and operating conditions at various locations throughout the piping system to the attributes and criteria in Table 2. The design and operating conditions are obtained from various sources, including plant operating procedures, design transient analyses, service experience, piping arrangement drawings, piping and instrumentation drawings, and isometric drawings. The piping and instrumentation drawings are used to identify the general layout of the piping system, while the arrangement drawings and isometrics identify the piping spatial configuration and weld locations. Generally, the degradation mechanism analysis is performed at each weld location in a pipe system.

5.2. FAILURE MODES OF PIPING

The modes (or types) of pipe failures depend on the conjoint requirements for material degradation, and they include axial or circumferential cracking (non-through-wall and through-wall), wall thinning, pinhole size leaks and axial splits. Large water hammer pressure transients can result in a complete (100%) ductile pipe rupture. For cracks, the crack propagation path can be circumferential, axial or they can take on a curvilinear relationship (e.g. initially circumferential and eventually assuming a mostly axial direction). Figure 25 gives examples of relationships between pipe failure mode and degradation mechanism.

Determining the possible consequences (i.e. geometry of through-wall cracking and the resulting mass flow rate) of a degraded condition is not a straightforward problem. While insights from OPEX and experimental work are valuable, in applying a piping reliability model some form of extrapolation is needed. However, this scheme of relationships enables one to identify relevant mechanisms at a specific location, or in turn identify and count specific failure-relevant locations within a plant system. This is the starting point for data driven or mechanistic analyses of expected failure rates. The national codes and standards for operability determination and fitness for service analysis provide further details on the determination of pipe failure mode with respect to the analysis method for flawed pipes. Depending on the piping material and method of fabrication or installation, and on an analysis method for determining fitness for continued operation, a limit load is controlled by plastic collapse, elastic-plastic fracture mechanics or linear elastic fracture mechanics.

5.3. DEGRADATION MECHANISMS APPLICABLE TO ADVANCED WCRs

This section addresses the dominant material degradation mechanisms that are applicable to an advanced WCR and how to determine whether projected degradation mechanisms would be different from those experienced in the WCR operating environments. These are complex questions that have received extensive coverage in numerous R&D efforts [75–81]. Assessments of the conjoint requirements for material degradation in advanced WCRs should be based on a formal degradation mechanism analysis that acknowledges the current material science state of knowledge as well as any relevant OPEX insights.

TABLE 2. SUMMARY OF FIRST IN-KIND MATERIAL DEGRADATION DISCOVERIES

Material degradation mechanism		Time period when first discovered in service							Notes/references
		1965–1969	1970–1974	1975–1979	1980–1984	1985 – 1989	1990–1994	1995–1999	
Corrosion mechanisms	Corrosion — general	☒	☐	☐	☐	☐	☐	☐	Widespread in NPPs but subjected to monitoring programmes. Susceptible systems include raw water piping systems and below ground/buried piping systems
	Corrosion under Insulation	☐	☒	☐	☐	☐	☐	☐	A recognized problem for more than 60 years. The first corrosion under insulation related standard was issued in 1971; ASTM-C692
	Corrosion — crevice	☐		☒	☐	☐	☐	☐	Both forms of material degradation were identified after a few years after start of commercial operation
	Corrosion —pitting	☐	☒	☐	☐	☐	☐	☐	
	Dealloying — selective leaching	☐	☐	☐	☐	☒	☐	☐	Some plants use Al-bronze piping in raw water service. Selective leaching was identified at these plants in the mid-1980s
	Galvanic corrosion	☒	☐	☐	☐	☐	☐	☐	'Bi-metallic' corrosion has been recognized for hundreds of years. Some plants continue to experience this form of degradation where partial piping replacements are made. For example, welding of high alloy stainless steel to carbon steel flange without installing galvanic protection

TABLE 2. SUMMARY OF FIRST IN-KIND MATERIAL DEGRADATION DISCOVERIES (cont.)

Material degradation mechanism		Time period when first discovered in service							Notes/references
		1965–1969	1970–1974	1975–1979	1980–1984	1985 – 1989	1990–1994	1995–1999	
	Graphitic corrosion	☐	☐	☐	☒	☐	☐	☐	Beaver Valley Unit 1 (PWR) experienced a fire water system pipe break on 8 Sep. 1982. The failure occurred in a section of below ground cast iron (ANSI 21.6) piping
	Microbiologically influenced corrosion	☐	☒	☐	☐	☐	☐	☐	Initial academic research about microbiologically influenced corrosion started in the 1930s. Many NPPs that were commissioned in the late 1960s experienced microbiologically influenced corrosion failures after a few years in commercial operation
	Erosion	☐	☒	☐	☐	☐	☐	☐	Most if not all commercial WCRs have experienced the three forms of erosion mechanisms. Erosion and erosion-corrosion is a problem in carbon steel piping. Erosion-cavitation can degrade carbon steel and stainless steel
	Erosion-cavitation	☐	☒	☐	☐	☐	☐	☐	
	Erosion-corrosion	☐	☒	☐	☐	☐	☐	☐	
Flow assisted degradation	Flow accelerated corrosion	☒	☐	☐	☐	☐	☐	☐	Big Rock Point (BWR) experienced several extraction steam flow accelerated corrosion failures in the 1966 to 1970 time frame (ORNL/NSIC-202). Another early flow accelerated corrosion event was recorded in 1971 when a DN100 elbow in the HPCI system failed at Dresden-2 (BWR)

TABLE 2. SUMMARY OF FIRST IN-KIND MATERIAL DEGRADATION DISCOVERIES (cont.)

Material degradation mechanism		Time period when first discovered in service							Notes/references
		1965–1969	1970–1974	1975–1979	1980–1984	1985 – 1989	1990–1994	1995–1999	
	Flow accelerated corrosion — entrance effect (leading edge effect)	☐	☐	☐	☐	☒	☐	☒	First reported in the 1990s
	Liquid droplet impingement erosion	☐	☐	☐	☒	☐	☐	☐	Salem Unit 2 (PWR) experienced a liquid droplet impingement erosion failure on 7 Jul. 1983
	Corrosion fatigue	☐	☒	☐	☐	☐	☐	☐	First reported for La Crosse demonstration BWR in 1970; Proc. Investigation of Corrosion Failures in Nuclear Reactor Component, III Inter-American Conference on Materials Technology 14–17 Aug. 1972
Fatigue									
	High cycle fatigue	☒	☐	☐	☐	☐	☐	☐	A high cycle fatigue failure was reported on 23 Feb. 1967 at the German plant KRB-I-A (Gundremmingen-A); a dual-cycle BWR
	Low cycle fatigue	☐	☒	☐	☐	☐	☐	☐	Robinson-2 (PWR) reported a low cycle fatigue failure on 20 Dec. 1972
	Thermal stratification, cycling and striping	☒	☐	☐	☐	☐	☐	☐	The Ågesta heavy water reactor in Sweden experienced a thermal fatigue failure of a mixing tee in 1966. The reactor was in operation between 1964 and 1974

TABLE 2. SUMMARY OF FIRST IN-KIND MATERIAL DEGRADATION DISCOVERIES (cont.)

Material degradation mechanism		Time period when first discovered in service							Notes/references
		1965–1969	1970–1974	1975–1979	1980–1984	1985 – 1989	1990–1994	1995–1999	
	External chloride induced SCC	☐	☐	☒	☐	☐	☐	☐	Several German plants experienced ECSCC of small diameter stainless steel lines in the 1975–1985 time frame; NEA/CSNI/R(2010)15
Stress corrosion cracking	Intergranular SCC	☒	☐	☐	☐	☐	☐	☐	First reported in 1965. Several experimental and prototype reactors experienced intergranular SCC prior to 1965
	Primary water SCC — Alloy 600/82/182	☐	☒	☐	☒	☐	☐	☐	Primary water SCC of steam generator tubes was first detected in Germany 1972. Cracking in thick section wrought Alloy 600 nozzles was first detected in PWR pressurizer nozzles in the 1980s
	Strain induced corrosion cracking	☐	☐	☒	☐	☐	☒	☐	Reported in Germany in the 1975–1985 time frame
	Transgranular SCC	☒	☐	☐	☐	☐	☐	☐	Several experimental and prototype reactors experienced transgranular SCC in the 1960s

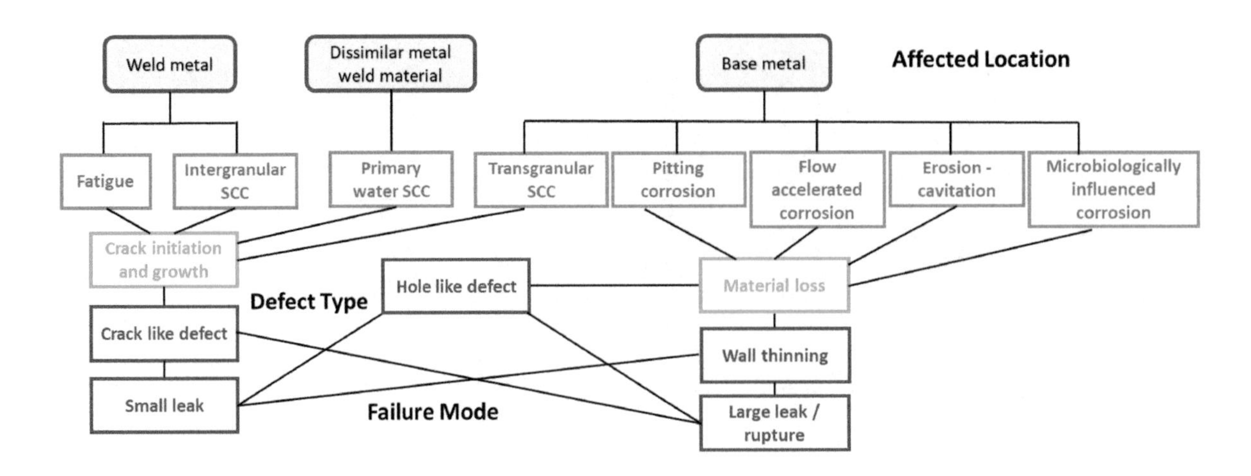

FIG. 25. Relationships between pipe failure mode and degradation mechanism.

The principal WCR piping materials are: austenitic stainless steels, cast-austenitic stainless steels, nickel base alloys, low alloy steels and carbon steels. The same basic material types are being used in advanced WCRs. Summarized in Table 4 are the known WCR material degradation mechanisms and when they were first discovered. This evaluation is based on the international OPEX. No new degradation mechanisms have been discovered in the past two decades. Some degradation mechanisms have been discovered in locations originally assumed to be immune, however. Also, some new materials have been developed by steel producers and subsequently applied as replacements for the original as-installed piping material. Examples includes Alloy 690 and super-austenitic (or high alloy) stainless steels. Both types were originally developed in the 1970s but with relatively limited application in the commercial nuclear industry until the 1980s. The WCR material degradation mechanisms also apply to the advanced WCRs. What matters the most in determining the possible severity of these mechanisms to advanced WCR operating environments is the effectiveness of the mitigation practices and the quality of RIM programmes that are implemented to minimize leakages. This is also a consequence of the 'Plan-Do-Check-Act' cycle for the improvement of the ageing mechanism as described by the IAEA Safety Guide SSG-48 [82]. In other words, the factor of improvement that can be achieved in material degradation susceptibility is to be established and applied appropriately in the piping reliability models.

The factor of improvement reflects the reduction in a mechanistic degradation property, such as crack growth rate, initiation frequency or fatigue endurance. Some examples are provided as follows:

— Carbon steels are extensively present in the balance of plant piping systems such as in the high and low pressure elements of the main steam system, extraction steam system, condensate, feedwater, feedwater heater drains and vents piping. Flow assisted degradation, mainly flow accelerated corrosion, caused extensive pipe failures in the 1970s and 1980s. Replacing the flow accelerated corrosion susceptible carbon steel with low alloy steel or stainless steel has resulted in a factor of improvement over 10 times with respect to material degradation resistance [83].
— Alloy 600 is used extensively in PWR NPPs. The material science community determined in the 1960s that this alloy was susceptible to SCC in certain high temperature environments. The first primary water SCC events in commercial PWR NPPs were reported in the mid-1980s (Table 2). Alloy 690 was determined to be the preferential replacement material. To date there are no known Alloy 690 failures and the factor of improvement (with respect to crack growth rate) is considered to be much greater than 10 [84].
— Carbon steels are used extensively in raw water cooling systems. The rate of corrosion failures has remained high where carbon steel piping is still in use. High alloy austenitic stainless steels were

especially developed for corrosive environments and were first introduced as replacement material for service water piping in the early 1980s. The high alloy stainless steel has resulted in a factor of improvement >>10 with respect to material degradation resistance [85].

Factor of improvement measures can be determined on the basis of OPEX data, experimental test data and/or expert elicitation. A strong technical basis does exist for adjusting WCR pipe failure rates using factor of improvements to account for the projected improvements in advanced WCR piping pressure boundary integrity.

6. EXTRACT AND SCREEN OPERATING EXPERIENCE DATA

6.1. OBJECTIVES

Reviews of OPEX with piping systems have been ongoing ever since the first commercial NPPs came on-line in the 1960s [1, 3]. In 1975, the US NRC established a first Pipe Crack Study Group charged with the task of evaluating the significance of SCC in BWRs and PWRs [80, 81]. Extracting, screening and analysing this OPEX were key activities of the work by the Pipe Crack Study Group. As another example, major condensate and feedwater piping failures due to flow accelerated corrosion resulted in national and international initiatives similar to that of the Pipe Crack Study Group [84–87].

Lessons learned through systematic evaluations of OPEX data were important inputs to the development of mitigation strategies to prevent recurrence of pipe failures. Irrespective of the technical approach, some form of OPEX data is used either implicitly or explicitly to inform or validate the models and/or the results. The process of extracting and screening OPEX data involves either developing a new database or accessing an existing data resource and to apply database query statements according to the reliability model input data needs. A query is a request for certain event populations, for action on data, or for both. It may be defined by, for example, assessing how many socket weld failures occurred in branch connections in PWR volume control systems during calendar years 2010 through 2020.

6.2. DATA QUALITY

The usefulness of any component failure data collection depends on the way by which a stated purpose is translated into database design specifications and requirements for data input and validation, access rules, support and maintenance, and quality assurance. In order to meet the data quality objectives a coding format is developed and it is usually documented in the coding guideline. It builds on established pipe failure data analysis practices and routines that acknowledge the unique aspects of passive component reliability in heavy water reactor and light water reactor operating environments (e.g. influences by material properties, water chemistry, temperature, pressure). For an event to be considered for inclusion in the event database, it undergoes an initial screening with an objective to go beyond the abstracts of event reports ensuring that only events according to the work scope definition are included in the database. This screening process sometimes is not straightforward. As one example, a PWR unit in 2016 experienced what initially appeared to be a minor reactor coolant pressure boundary leakage on a high pressure safety injection line. On closer evaluation, the leak was located on a seal weld of a threaded small diameter connection and the leakage path was via the threads and not through the pipe wall. Therefore, the leakage was not a reactor coolant pressure boundary leakage per ASME XI definition. Subsequently this event was not selected for inclusion in the database.

Data quality is affected as soon as the field experience data is recorded at an NPP, interpreted, and then entered into a database system. The field experience data is recorded in different types of information systems such as the action requests, work order systems, via ISI databases and outage summary reports, as well as the licensee event reports or reportable occurrence reports. Therefore, the details of a degradation event or failure tend to be documented to various levels of technical detail. Building a database record containing a full event history frequently involves extracting information from several sources. The term 'data quality' is an attribute of the processes that have been implemented to ensure that any given database record (including all of its constituent elements, or database fields) can be traced to the source information. The term also encompasses fitness for use, meaning that the database records should contain sufficient technical detail to support database applications.

6.3. SOURCES OF PIPE FAILURE DATA

Examples of pipe failure data resources are summarized in Annex III. The data resources are of three different types:

- *Type 1.* A database developed especially for piping reliability analysis tasks. Based on a detailed piping reliability taxonomy, which may consist of more than 800 data filters (or key words) in order to support data extraction, screening and analysis. An example of such a taxonomy is found in [92].
- *Type 2.* General purpose OPEX database from which information about pipe failures can be extracted. An example is the NEA (Nuclear Energy Agency)/IAEA Incident Reporting System database.
- *Type 3.* Piping reliability parameter databases (or handbooks) that include tabulations of pipe failure rates with uncertainty distributions.

6.4. DATABASE ACCESSIBILITY

Access to OPEX data is important to the performance of piping reliability analysis tasks. However, accessing the available OPEX data is potentially a complex matter since there are no open source pipe failure databases. While several databases exist that are dedicated to pipe degradation and failures, these databases are either restricted or proprietary and have been developed and maintained over long periods of time. An example of a restricted database is the OECD (Organisation for Economic Co-operation and Development)/NEA CODAP (Component Operational Experience, Degradation and Ageing Programme) database which has specific rules for data access [93]. CODAP is a web based SQL[7] database which was established in 2002. Details on how to gain access to this database can be found on the OECD NEA web site.

The completeness of databases refers to the completeness of the event specific technical information. It also refers to the completeness of the event populations so that there is an assurance that all relevant events are captured in a database. The processes and procedures for maintaining pipe failure databases vary significantly. Figure 26 shows examples of database infrastructure considerations versus possible expectations that may be imposed on database structure and content.

[7] Structured Query Language (SQL) is a domain specific language used in programming.

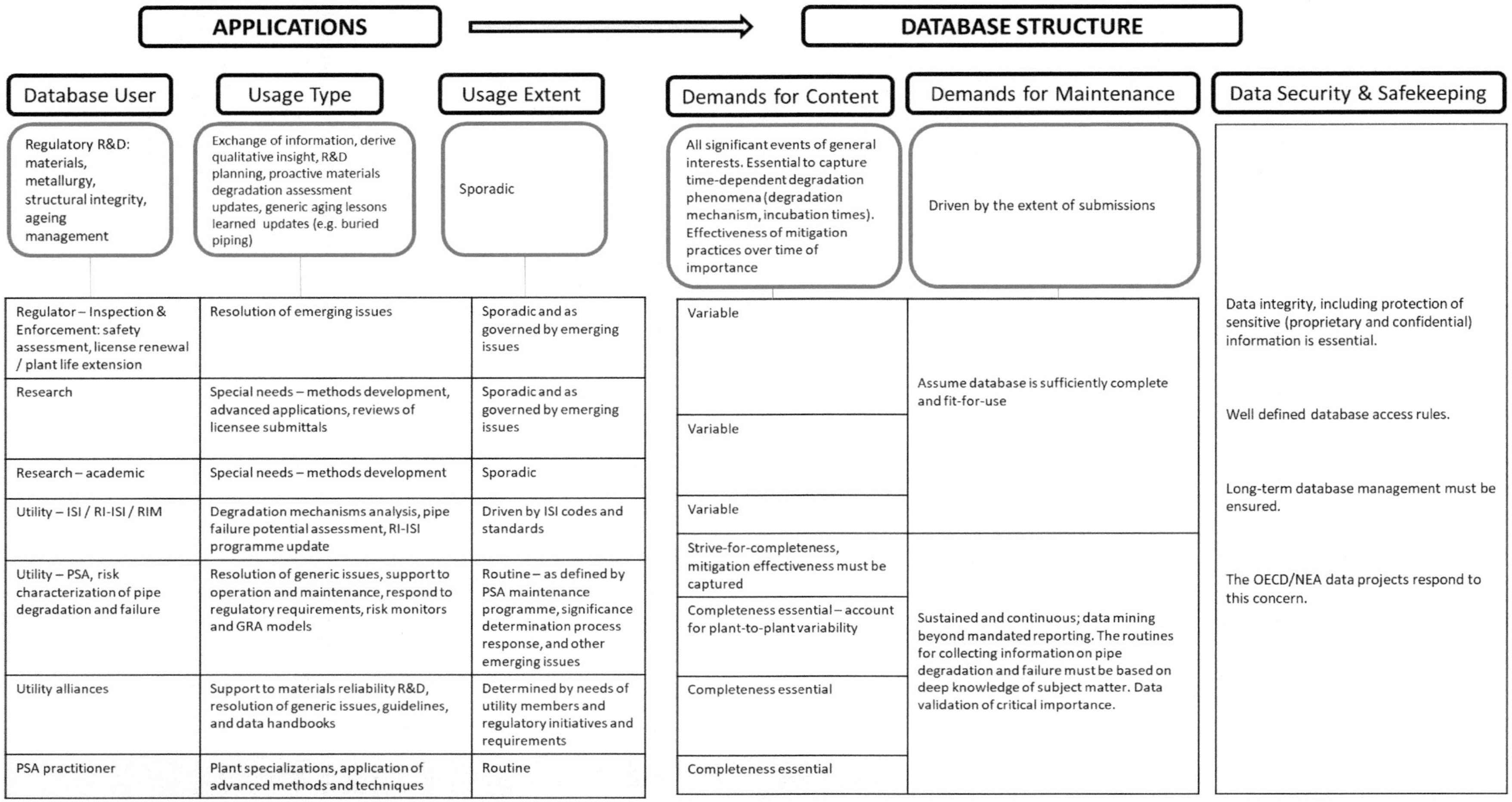

APPLICATIONS			DATABASE STRUCTURE		
Database User	**Usage Type**	**Usage Extent**	**Demands for Content**	**Demands for Maintenance**	**Data Security & Safekeeping**
Regulatory R&D: materials, metallurgy, structural integrity, ageing management	Exchange of information, derive qualitative insight, R&D planning, proactive materials degradation assessment updates, generic aging lessons learned updates (e.g. buried piping)	Sporadic	All significant events of general interests. Essential to capture time-dependent degradation phenomena (degradation mechanism, incubation times). Effectiveness of mitigation practices over time of importance	Driven by the extent of submissions	
Regulator – Inspection & Enforcement: safety assessment, license renewal / plant life extension	Resolution of emerging issues	Sporadic and as governed by emerging issues	Variable	Assume database is sufficiently complete and fit-for-use	Data integrity, including protection of sensitive (proprietary and confidential) information is essential. Well defined database access rules. Long-term database management must be ensured. The OECD/NEA data projects respond to this concern.
Research	Special needs – methods development, advanced applications, reviews of licensee submittals	Sporadic and as governed by emerging issues	Variable		
Research – academic	Special needs – methods development	Sporadic	Variable		
Utility – ISI / RI-ISI / RIM	Degradation mechanisms analysis, pipe failure potential assessment, RI-ISI programme update	Driven by ISI codes and standards	Strive-for-completeness, mitigation effectiveness must be captured		
Utility – PSA, risk characterization of pipe degradation and failure	Resolution of generic issues, support to operation and maintenance, respond to regulatory requirements, risk monitors and GRA models	Routine – as defined by PSA maintenance programme, significance determination process response, and other emerging issues	Completeness essential – account for plant-to-plant variability	Sustained and continuous; data mining beyond mandated reporting. The routines for collecting information on pipe degradation and failure must be based on deep knowledge of subject matter. Data validation of critical importance.	
Utility alliances	Support to materials reliability R&D, resolution of generic issues, guidelines, and data handbooks	Determined by needs of utility members and regulatory initiatives and requirements	Completeness essential		
PSA practitioner	Plant specializations, application of advanced methods and techniques	Routine	Completeness essential		

FIG. 26. Database infrastructure requirements vs. applications.

7. METHOD SELECTION

7.1. THE CONTEXTS AND DOMAINS OF PIPING RELIABILITY ANALYSIS

The selection and application of a piping reliability methodology is conditioned by the objective of an analysis and the domain in which the outcomes of an analysis are to be applied. It is a three-tiered selection and application process that acknowledges the general context of an analysis, end user expectations and requirements, and relevant national codes and standards. Table 3 gives examples of three different piping reliability analysis contexts: (1) PSA including design certification PSA of new reactor designs, (2) codes and standards, and (3) risk-informed operability determination and probabilistic fitness for service assessment. Table 4 gives examples of different pipe failure rate analysis requirements. Both tables apply to operating advanced WCRs as well as the evolving (conceptual to final) design phases of a proposed advanced WCR.

TABLE 3. EXAMPLES OF PIPING RELIABILITY ANALYSIS CONTEXTS

Objective(s)	Analysis objective	Description
PSA, PSA applications and design certification PSA	LOCA initiating event frequencies	Quantification of location specific LOCA frequencies as a function of EBS or through-wall flow rate. The analysis addresses multiple sensitivity cases to account for different leak detection and RIM strategies
	High energy line break initiating event frequencies	Quantification of location specific high energy line break frequencies inside and outside the containment. The analysis addresses safety related and non-safety related high energy line break scenarios
	Internal flooding initiating event frequencies (internal flooding PSA)	Quantification of internal flooding scenarios attributed to pipe failure. Pipe failure frequencies account for all potential flood sources; safety related and non-safety related piping, including fire water piping system
	Support system failure initiating event frequencies	Examples include loss of cooling (open and closed loop cooling systems), loss of instrument air and failure of electro-hydraulic control system
	Estimation of pipe failure frequency due to hydrogen deflagration	NPPs use hydrogen to cool turbine generators and also to condition the primary circuit coolant
RIM per ASME XI Division 2 (2019) or equivalent code	RIM programme development for a new advanced WCR	Quantification of reliability targets for different ISI locations
	RIM programme development for an existing advanced WCR	Same as for 'new advanced WCR' except that updates may have to be performed to account for new OPEX or design modifications

TABLE 3. EXAMPLES OF PIPING RELIABILITY ANALYSIS CONTEXTS (cont.)

Objective(s)	Analysis objective	Description
Operability determination or fitness for service	Risk-informed 'significance determination'	Application of plant specific PSA to determine the risk significance of a degraded or failed piping component. The analysis involves quantifying piping reliability prior to the discovery of a degraded/failed component as well as quantifying piping reliability in light of the new OPEX
	Structural evaluation of flawed piping component	An analysis based on probabilistic structural reliability modelling is performed to determine whether a flawed component is suitable for continued operation (e.g. until next planned outage of sufficient length to accommodate a repair or replacement)

TABLE 4. EXAMPLES OF PIPING RELIABILITY ANALYSIS REQUIREMENTS

PSA task	Description	Piping reliability modelling requirements
Initiating event frequency	LOCA due to primary coolant pipe failure	The exact definition of LOCA is a function of plant design, including emergency core cooling system make-up capability. Engineering calculations support the plant specific classification of the different sets of LOCA frequencies that are needed. The LOCA frequency calculation accounts for degradation mechanisms and all the locations within the primary pressure boundary that potentially could produce a LOCA of certain magnitude. The PSA model input can be in the form of a cumulative pipe failure frequency versus the through-wall mass flow rate or EBS. The analysis acknowledges locations that were stress relieved during construction or at later dates to mitigate possible SCC effects. Consideration of RIM (e.g. leak detection and ISI) is an explicit analysis requirement
	Main steam line break inside and outside the containment Feedwater line break inside and outside the containment	Assessment of location specific pipe failure rates. Engineering analyses support the definition of calculation cases followed by detailed reviews of piping arrangement drawings and isometric drawings. The analysis addresses dynamic pipe break effects and the spatial impact of consequential actuation of fire protection sprinklers that results in full flow from fire protection system pumps into the turbine building
Internal flooding initiating event frequency	Assessment of all possible sources of internal flooding caused by pipe failure	Internal flooding PSA is concerned with pipe failure locations inside the reactor building, auxiliary building, service building and turbine building. Pipe failure rates are developed for multiple systems, pipe sizes, locations within a plant and operating environments

The domains of piping reliability analysis refer to the intended utilization of analysis results. There are three types of end users of results in support of: (1) regulations, (2) operations, and (3) reactor design (Fig. 27). In a regulatory domain a piping reliability analysis may be performed in response to new regulatory requirements, to make changes to an existing licensing basis, or to assess the risk significance of pipe failure. In an operational domain a piping reliability analysis may be performed to optimize ISI and ageing management programmes (e.g. adding or removing inspection locations), or to assess the risk-significance of a leaking piping component. Finally, in the reactor design domain a piping reliability

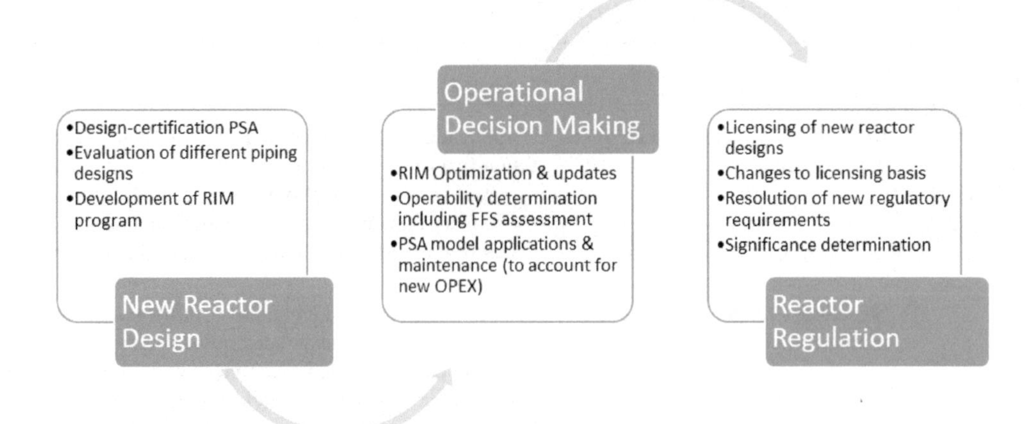

FIG. 27. Three domains of piping reliability analysis.

analysis would support the evaluation of different piping system design concepts or the development of an RIM programme.

Associated with the respective domain are different end user requirements. These requirements are found in regulations (high level), regulatory guides (specific), and codes and standards. The requirements provide details on the scope of validation of methods, inputs and outputs, documentation and peer review.

7.2. PIPING RELIABILITY METHODOLOGIES

The state of the art piping reliability methodologies fall into three general categories: (1) DDM, (2) PFM, and (3) I-PPoF, which incorporate aspects of DDMs and PFM in an integrated probabilistic physics-of-failure modelling approach. The three categories share common elements (Fig. 28) that account for what is known about structural integrity in different operating environments, through OPEX, experimental data and engineering analyses. The methodologies account for different types of materials, availability of data, and the effects of RIM on structural integrity, including the role of ISI, leak detection, stress improvement processes and ageing management.

Expert elicitation is sometimes used to develop additional qualitative and quantitative information on a specific technical aspect of the structural integrity of a certain piping configuration, [94]. The objective of an expert elicitation is to express a state of knowledge in terms of a probability distribution which is input to a piping reliability analysis. The implementation of a DDM, PFM or I-PPoF involves the

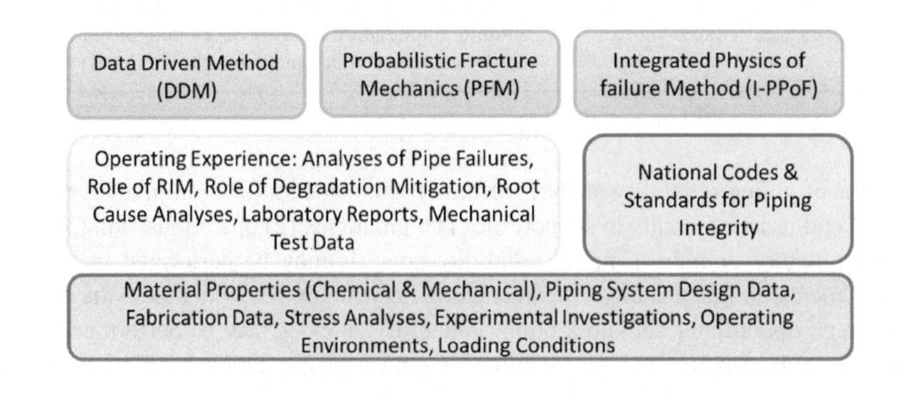

FIG. 28. Different piping reliability analysis methodologies and their common elements.

use of computational tools; open source and/or proprietary software tools. The practical implementation of a respective method tends to be computationally intensive and requires planning, quality assurance of input parameters, definition of sensitivity cases, results interpretation and detailed documentation to enable an independent review. For advanced WCRs, the availability of relevant data influences the methods selection process as well as the methods implementation.

Details of the three different methodologies are provided in Annex I. Each methodology may be applied synergistically whereby outputs from one method provide input to another method. For example, conditional failure probabilities calculated using PFM can be used as input to data driven models. Or output from data driven models can be used to calibrate PFM models against OPEX data. Sometimes expert judgement (i.e. through an expert elicitation process) is used to support an application by providing justifications for certain assumptions. When selecting methodology, the following considerations apply [95].

— *Modelling of material degradation.* Consideration of the physics of material degradation and failure can be implicit or explicit. An overly simplified statistical analysis of pipe failure would mask the relationships among the probabilistic failure metrics, physical failure mechanisms and underlying physical factors (e.g. material properties, geometry, operating environment). On the other hand, explicit consideration is obtained by using a physics based model of pipe flaw initiation and propagation.
— *Failure characterization.* DDMs estimate the probabilistic failure metrics on the basis of OPEX data and engineering analyses (e.g. degradation mechanism analysis, assessments of the effectiveness of degradation mitigation), with or without performing fracture mechanics analyses. In contrast, PFM and I-PPoF estimate the probabilistic failure metrics by comparing the cumulative damage (as a function of time) with the endurance level of the component being analysed.
— *Types of physical models.* Correlation based models rely on engineering correlations derived by fitting an analytical functional form to the experimental data. Mechanistic models numerically solve governing equations derived from theoretical laws of physics (e.g. conservation laws consisting of continuity, momentum conservation and energy balance). An integration of correlation based and mechanistic models utilizes the combination of correlation based and mechanistic models.
— *Modelling of RIM processes.* The reviewed studies under no/implicit consideration of RIM processes in Fig. 29 either did not address the influence of RIM (e.g. only physical degradation is modelled) or implicitly considered the RIM impact through statistical analyses of empirical data collected from repairable systems without separating the RIM impact from the physical degradation impact. In the studies under explicit incorporation of RIM processes, the relationship between the probabilistic failure metrics and the RIM parameters such as the time to repair, probability of detection and inspection intervals, was captured explicitly through a model.

The DDM is implicit with respect to the incorporation of physics-of-failure, while the PFM and I-PPoF explicitly incorporate the physicality aspect such as crack initiation and crack propagation. As described in this publication, the DDM includes an explicit consideration of RIM processes since a Markov model extension is used to compute the impact of different RIM strategies. Therefore, the DDM methodology can be considered as category (A.1.2) in Fig. 29. The DDM has a commonality with the I-PPoF methodology in terms of the nature of the coupling between physical degradation and RIM processes as both methods use a renewal process model for the coupling; however, they differ in whether the incorporation of physics is implicit vs. explicit.

Both PFM and I-PPoF use the damage-endurance analysis for failure characterization and the integration of correlation based models and mechanistic models for physical degradation modelling. Associated with criterion #4, both PFM and I-PPoF explicitly consider RIM processes. Therefore, PFM and I-PPoF can be considered as category (B.2.3.2) in Fig. 29. Yet, comparing PFM and I-PPoF, there are three key differences worth noting. First, the I-PPoF methodology combines finite element analysis for detailed thermo-mechanical analysis with crack initiation and crack propagation models, while the PFM

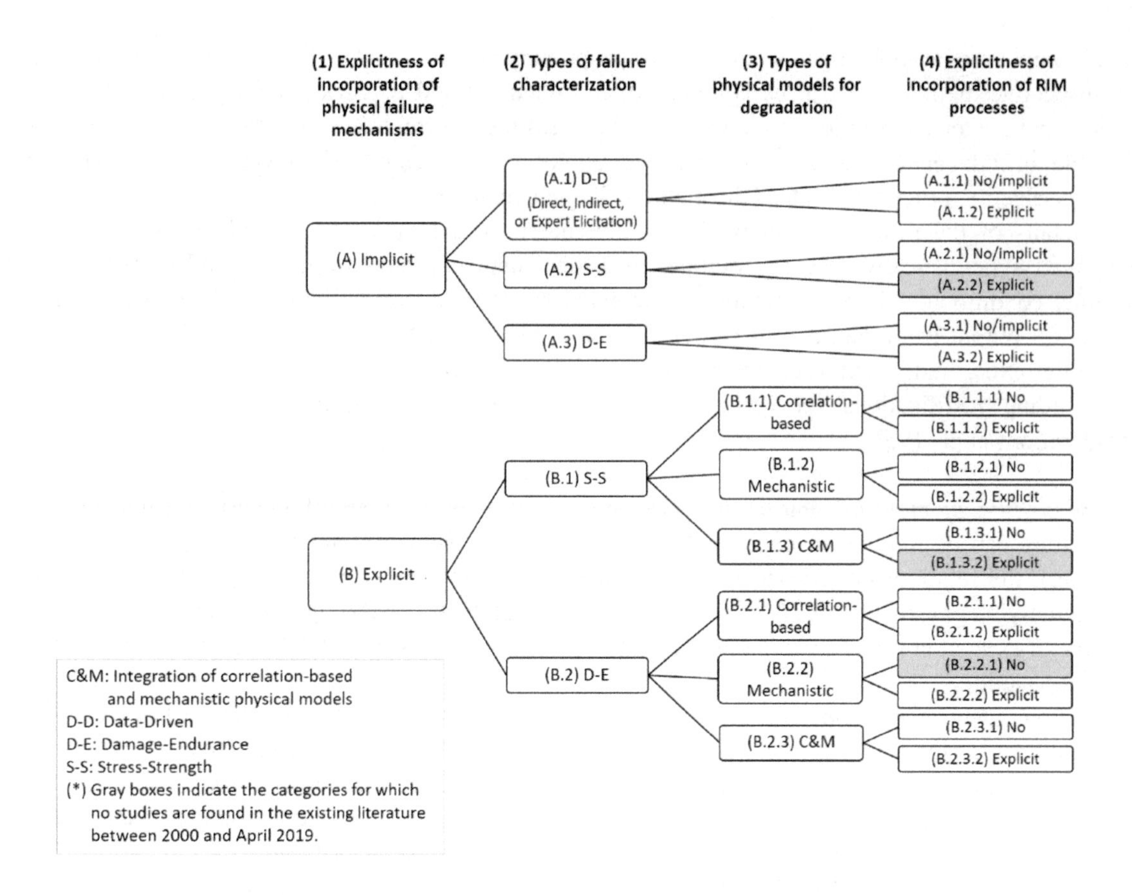

FIG. 29. Categorization of studies on the estimation of probabilistic failure metrics.

methodology is able to use validated mechanistic models and computing efficient analytical solutions. Second, the modelling of interactions between physical degradation and RIM processes differs. PFM mainly uses discrete event simulation, where degradation and failure mechanisms are simulated over time using physics based models, and the RIM processes such as ISI and leak detection are considered when the simulation time or physical condition reaches the predefined threshold. In contrast, I-PPoF uses renewal process models to account for the interactions between physical degradation and RIM processes, where the degradation transition rates are estimated from physics based models while the RIM transition rates are estimated from an RIM work process model. Third, the I-PPoF method accounts for the effects of different crack depths and the time for crack growth, by decreasing the number of surface crack stages to one. This approach implies that the crucial interaction of crack growth and inspection intervals, which is a key question for PFM, cannot be covered in the I-PPoF approach.

7.3. APPLICABILITY OF THE DIFFERENT METHODOLOGIES

The question on how to select an appropriate methodology is not strictly technical regarding validity and peer recognition, availability of data and computational tools, to mention some. It also relates to the resource requirements for implementation such as, but not limited to, tools development, training of analysts, project schedule constraints and data availability. The results and analysis insights have to be fit for use and the underlying documentation should be of sufficient quality to facilitate an independent peer review. Drawing on practical analysis insights and results, an assessment of the feasibility of applying a certain methodology to a certain type of problem is given in Table 5.

TABLE 5. THE RANGES OF APPLICABILITY

Application category	Intended application	Feasibility of piping reliability methodology		
		DDM	PFM	I-PPoF
Probabilistic safety assessment (PSA)	Safety class 1 piping	High	High	(High)
	Safety class 2 piping	High	High	TBD
	Safety class 3 piping	High	Low	TBD
	Balance of plant piping	High	Low	TBD
	Pipe failure locations for which stress analysis reports are available	n.a.	High	TBD
RIM programme development and programme maintenance	Definition and prioritization of ISI locations	High	High	TBD
	Update of RIM programme to address new OPEX	High	High	TBD
	Optimization of inspection locations by accounting for leak detection and probability of detection	High	High	TBD
	ISI/Codes and Standards Relief Request Support			
	Impact of deferred inspection one or more inspection intervals	High	High	TBD
	Impact of deferred hydrostatic pressure testing	High	High	TBD
	Acceptability of temporary repair vs. codes and standards repair	n.a.	High	TBD
Codes and standards and risk informed regulation	Demonstration of 'very low' probability of failure — maximum allowable CFP	High	High	TBD
	Leak before break applications	Moderate or n.a.	High	TBD
	Fitness for service assessment	n.a.	High	TBD
	Flaw tolerance analysis	n.a.	High	TBD
	ASME XI code case applications	n.a.	High	TBD
	Risk significance determination	High	Moderate	Moderate

Note: n.a. — not applicable.

Different formulations of DDMs have been used since the dawn of applied risk and reliability analysis. These early formulations relied on non-nuclear industry data on pipe failures and limited WCR data [1, 3, 4, 12]. The PFM has its roots in theoretical studies that were performed in the 1970s [96]. Physics-of-failure models of reliability were first proposed in the early 1960s. However, application of physics-of-failure concepts in structural reliability is a relatively recent development [97].

Table 5 provides an implied reference to the current, state of the art implementations of the respective modelling concept. Each concept has evolved significantly in recent years. The following nomenclature is used:

— *Feasibility*. Recognized through the combination of peer review, extent of user experience, availability of data and computational tools, and in recognition of resource constraints to produce realistic results. It is also concerned with the feasibility of successfully applying a methodology by an analyst that has not been directly associated with the methods development:
  - High. Demonstrated to be relevant to a given analytical context. An application can be completed within the constraints of a given project schedule.
  - Moderate. In principle, the methodology is relevant, but its feasibility is contingent on having highly experienced analysts in order to produce meaningful results.
  - Low. Applicability of the methodology is unproven or could be technically challenging, requiring additional R&D.
  - TBD (to be determined or unproven). Primarily, the methodology has been used in an academic or R&D domain.
  - n.a. (not applicable). The methodology is not applicable or is not intended for a certain type of application (e.g. fitness for service evaluation in support of an ISI relief request).

The following general findings can be drawn from benchmark studies such as the ones described in [24], the risk-informed ISI methodology benchmark [98], numerous PFM benchmark case studies including nuclear risk based inspection methodology [99, 100], as well as from the broader field of NPP applied risk and reliability analysis [101]:

— *DDM feasibility*. Demonstrated as a feasible approach for assessing location specific pipe reliability parameters and for a relatively complete set of advanced WCR piping systems (both safety and non-safety related), degradation mechanisms and operating environments. However, the DDM applications require computational techniques and tools especially developed for piping reliability assessments. The DDM generates probability density functions of the main variables and a total probability density function which describes the mode of pipe failure and a spectrum of consequences of failure (e.g. in terms of through-wall flow rates). Access to OPEX data is an intrinsic and essential aspect of the methodology. The DDM has been used extensively in support of PSA, design certification PSA of advanced reactors, internal flooding PSA, risk-informed ISI development, RIM programme development for advanced reactors and risk-informed operability determination.

— *PFM feasibility*. Demonstrated as a feasible approach for assessing location specific pipe reliability parameters for primarily safety class 1 piping system locations susceptible to SCC or fatigue mechanisms. The PFM computer code development continues to be supported by national and international R&D and benchmark studies. A strength of PFM is its ability to support parametric studies to determine the effect of, for example, assumptions about weld residual stress properties, crack growth rates and material properties. The PFM method generates probability density functions of the main variables and a total probability density function which describes the mode of failure and the consequences of failure. The PFM method has been used in risk-informed ISI programme development and it is frequently used in support of fitness for service evaluations. A summary of the current status of PFM is found in [102]. In the United States of America (USA), a regulatory guide for PFM applications has been issued to aid licensees in preparing submittals of PFM calculation results [103]. A regulatory review process would include the process for establishing model input parameters, software tool validation, and the extent by which the results reflect observations from operational experience evaluations [104].

— *I-PPoF feasibility*. This approach has been tested in R&D environments. To simulate pipe degradation and failure, analysis procedures are developed and implemented for each degradation mechanism to propagate a flaw under prevailing loading conditions to predict the pipe failure frequency. An implementation of the I-PPoF approach involves an integration of multiple modules, including the use of finite element methodology for the determination of stress distribution parameters, ASTER, COMSOL or MATLAB computing environments for crack propagation evaluation, a human reliability analysis based model for the RIM activities (e.g. ISI, leak detection, repair/replacement), and renewal process models for quantification of state transition frequencies (crack initiation to crack, growth to leakage) accounting for interactions between the physical degradation and the RIM activities. Different types of multi-physics computational environments have been applied to piping reliability (e.g. [105–110]). Exploratory applications have been in areas such as ageing PSA and design certification PSA as well as for modelling of specific material degradation mechanisms. In theory, the I-PPoF approach applies to all combinations of structural materials and operating environments.

The additional details on comparative analyses and benchmark case studies are found, for example, in [111–113].

7.4. METHODOLOGY SELECTION CRITERIA

Before initiating an analysis, several factors need to be considered that will guide the selection of a methodology and how to structure an implementation which is appropriate to the needs of an analysis. Careful planning will help to ensure that the outputs ultimately fit the needs and are appropriate to the intended audience, and that the analysis can be reasonably completed within resource constraints (e.g. schedule, data availability, analyst experience levels).

The selection criteria to be considered include: (a) how to organize an analysis and to define the requirements to be placed on an analysis task, (b) requirements with respect to analysis tools to be used, and (c) methodological considerations. High level requirements are included in Table 6 and these requirements reflect the following aspects of piping reliability analysis:

(1) *Analysis specification*. This was discussed in Sections 3, 4 and 5 of this report. How well does a certain methodology apply to a given evaluation boundary? Multiple evaluation boundaries will have to be addressed when performing a system level analysis (e.g. PSA initiating event frequencies that are attributed to a pipe failure). End user requirements determine the format for documenting the results. End user categories include:
   — PSA analysts performing initiating event frequency assessments and internal flooding risk assessments;
   — Regulatory staff performing reviews of fitness for service assessments;
   — Plant operators performing fracture mechanics evaluations of the integrity of a piping component with an embedded flaw (e.g. non-surface connected crack).

(2) *Complexity* of the problem to be solved.

(3) *Skill levels* and training needed. How easy is it to implement a certain methodology?

(4) *Transparency* of the analysis flow and ease of peer review.

(5) *Methodology implementation* requirements (e.g. special tools needed).

(6) *Validity*. Has the methodology been peer reviewed and applied to similar problems?

(7) *Input data* requirements, and resources needed for data processing and interpretation. Multiple sets of input data are required and supporting engineering analyses may have to be performed.

The implementation of a methodology takes on many different forms. Also, different analysis techniques and tools are applied in the implementation of the respective methodology. As an example, with respect to applications of PFM, the following requirements and approaches are proposed [99]:

Requirements:

(1) The PFM theory and technical basis to be published and independently reviewed.
(2) The PFM model and the associated software to address the relevant degradation mechanisms.
(3) The PFM model and the associated software to be able to assess failure leak event and rupture frequencies.
(4) Sensitivity study using the PFM model and the associated software to be presented, addressing relevant degradation mechanism with failure probabilities for small leaks to ruptures to be evaluated for variations of input parameters and to be consistent with expectations and the PFM theory assumptions.
(5) Sample calculations to be presented where the assigned input parameters are described and sources of the data assignments provided. The probability distributions and internally assigned parameters in the PFM software have to be documented together with the limitations of the PFM software.

Approaches:

(1) The PFM software to be benchmarked and report prepared and independently reviewed for the relevant material degradation mechanism against at least one other open source PFM software.
(2) The PFM software to be benchmarked against OPEX using actual plant failure frequencies. For damage mechanisms with no ruptures, leak frequencies may be used for the comparison.
(3) Hardwired formulations to be avoided as well as the risk of software misuse. Formulations of input data (type of probability distribution and its random properties) to be decided by the user giving a possibility of using a deterministic input for the variables.
(4) The PFM software to enable the user to extract control variables from the results such that a user is able to check the solutions and understand the results. Examples of control variables for a cracking damage mechanism are:
    — Initial crack sizes, crack shapes during the subcritical growth and critical crack sizes at leak and rupture;
    — Time to leak and rupture;
    — Stress intensity factors and J-integrals;
    — Crack opening areas and leak flow rates for through-wall cracks.
(5) The influence of inspections to be included in the structural reliability models and the associated software in order to quantify risk reductions from repeated inspections.
(6) Leak before break events to be modelled when considering and evaluating the rupture probabilities. In this context an adequate model of crack opening areas, leak flow rates and leak flow rate detection are significant.
(7) The used software to be clearly identified. It is desired that new information or better modelling assumptions be continuously incorporated into the structural reliability models and the associated software so that the generated results may replicate the best current knowledge.

The NUREG-1829 expert elicitation project [23] included a limited exercise to reconcile predicted weld failure rates with the reported service experience. The failure mode considered was perceptible leakage (e.g. visible moisture on the outside diameter pipe wall). PFM calculations using the WinPRAISE computer code generated predictions about the weld failure rate for different assumptions about the normal operating stresses (σ_{NO}). A DDM was used to derive weld failure rates from OPEX data, shown in Fig. 30. The controlling parameters (e.g. weld residual stresses) of the PFM model were calibrated

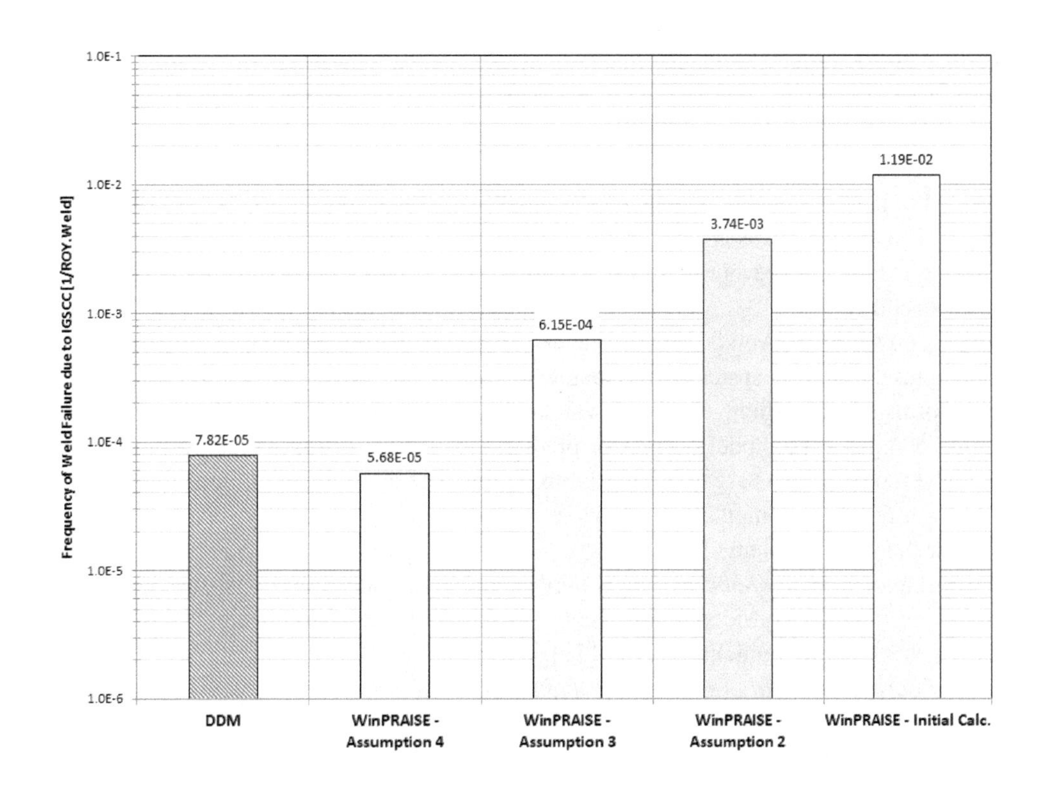

FIG. 30. Results of a DDM/PFM reconciliation task.

against the DDM results, responding to the question of how to reach agreement between the results of two different approaches to the estimation of pipe failure frequency.

The methodology selection criteria are closely linked to the context of an analysis and the end user domain, including the expectations that are placed on it (i.e. expected outcomes, results, insights). Itemized below are some of the modelling issues to be considered in the selection of an appropriate methodology as presented in Table 6:

— *Explicit versus implicit incorporation of physics-of-failure.* The controlling parameters of most piping material degradation mechanisms are well understood. As an example, the susceptibility of piping to flow accelerated corrosion is controlled by: (a) temperature, (b) flow conditions (e.g. single- or two-phase flow, turbulent versus laminar flow), (c) fluid pH value, (d) fluid oxygen content, (e) alloy additions, and (f) piping geometry. In view of the very extensive OPEX data, would it be beneficial to develop a physics-of-failure model of flow accelerated corrosion as part of the piping reliability modelling effort? Regardless of the chosen methodology, physics-of-failure is always implicitly accounted for in piping reliability analysis.

— *If physics-of-failure is explicit, the question is, what is the degree of realism in physics-of-failure?* This becomes an issue for known forms of material degradation for which limited/no OPEX data are available. Such an example is thermal ageing embrittlement of cast austenitic stainless steel piping components such as elbows. In a high temperature operating environment thermal ageing embrittlement causes a reduction in fracture toughness. Extensive research on thermal ageing embrittlement has been conducted and this form of material degradation is monitored through ageing management programmes. This research has included mechanical testing of components that have been removed from WCRs after many years of operation.

— *RIM processes are explicitly incorporated.* Implicit consideration of RIM (e.g. leak detection, ISI, pipe stress improvement) is always within the scope of an analysis, regardless of selected methodology. Many practical applications require explicit consideration of impact on reliability from changes to an existing RIM process. The factors that tend to be the most important are method and frequency of leak detection, non-destructive examination reliability (i.e. probability of detection) and pipe stress improvement.

— *If RIM processes are explicitly incorporated, the question is, what is the degree of realism?* An analyst is expected to have a deep knowledge of RIM, national codes and standards, and RIM qualification protocols. The analyst needs also to be aware of the RIM failure experience data, including reasons why RIM has failed to prevent a failure. Reference [42] gives examples of RIM failures; how and why they occurred.
— *The validation approach.* Validation is a crucial component of the analysis to ensure that the developed models represent the real system behaviour with an adequate level of accuracy and realism. There are two forms of validation: empirical validation and computational benchmark. The empirical validation compares the model outputs or predictions with the empirical and observational data to check if the model outputs are consistent with reality. The computational benchmark compares the outputs from different computational models with each other to identify their discrepancies when applied for the same problem.
— *The integrated treatment of uncertainty is implemented.* The questions to ask are, what are the types of uncertainty that are considered and what is the quantification approach. The treatment is to be consistent with existing guidelines [114–121].
— *Number of parameters that are treated probabilistically (i.e. characterized by a probability density function).* This is a function of the computational tools that are used to implement a methodology. Developing the appropriate distribution parameters could require a significant effort including the performance of engineering analyses.

TABLE 6. HIGH LEVEL METHODOLOGY SELECTION CRITERIA

Selection criteria	Piping reliability methodology		
	DDM	PFM	I-PPoF
Analysis specification			
Ease of implementation assuming an experienced analyst	MODERATE TO COMPLEX	MODERATE TO COMPLEX	COMPLEX
Resources needed for input data preparation and supporting engineering analyses	HIGH	HIGH	HIGH
Availability of open access tools	YES [a]	YES [a]	YES [a]
Proprietary tools with need for training	YES	YES	YES
Detailed, integrated treatment of uncertainties	YES	(YES) [b]	YES
Time to implement methodology	MODERATE	MODERATE	HIGH
Needs for technical support	MODERATE	(HIGH) [c]	HIGH
Ease of technology transfer	MODERATE	MODERATE	MODERATE
Ease of results interpretation	MODERATE [d]	MODERATE [d]	MODERATE [d]

TABLE 6. HIGH LEVEL METHODOLOGY SELECTION CRITERIA (cont.)

Selection criteria	Piping reliability methodology		
	DDM	PFM	I-PPoF
Required user proficiency level/Skill levels			
Entry-level / early career engineer — with tutoring as applicable	(YES)	(YES)	(YES)
Practicing engineer with 10+ years of relevant industry experience (materials, structures, operations, PSA)	YES	YES	YES
Expert in risk and reliability and/or fracture mechanics	PREFERRED	PREFERRED	PREFERRED
PSA peer review endorsement			
Meets applicable PSA standards requirements	YES	--	--
Recipient of 'Good Practice' note	YES [e]	--	--
Specificity w.r.t. plant design, piping design	YES	YES	YES
Integrated treatment of uncertainty	YES	(YES) [b]	YES
Quality of documentation	YES	YES	YES
Level of realism of results	YES	YES	YES
Comparing results against acceptance criteria	YES	YES	YES
Analysis tools and documentation of implementation requirements			
Feasibility of own implementation of methodology	(HIGH)	MODERATE	MODERATE
Training tools/documentation	YES	YES	YES
Proprietary tools and license requirements	NO	YES	NO
Training advised	YES	YES	YES
Validity			
Does a selected methodology support all types of evaluation boundaries; beyond safety class 1 locations?	YES	NO	(YES)
Extent by which methodology has supported practical applications	HIGH	HIGH	LOW
OPEX reconciliation	n.a.	(YES)	(YES)
Input data requirements			
Material properties (chemical and mechanical)	(NO) [f]	YES	YES

TABLE 6. HIGH LEVEL METHODOLOGY SELECTION CRITERIA (cont.)

Selection criteria	Piping reliability methodology		
	DDM	PFM	I-PPoF
Pipe stress analysis reports	NO	YES	YES
P&ID, fabrication isometric drawings, ISI isometric drawings, stress analysis isometrics	YES	(YES)	(YES)
OPEX data	YES	(YES)	(YES)
Degradation mechanism analysis	YES	(YES)	(YES)

Note: n.a. — not applicable.

[a] With reference to DDMs, a user implementation is feasible using MS-Excel and open access tools (GoldSim, MatLab, etc. but usually with some limitations attached w.r.t. usage). With reference to PFM, examples include the Piping Reliability Analysis Including Seismic Events (PRAISE) and xLPR PFM codes.

[b] Deterministic input is/or can be used.

[c] A parenthesis means that if an application is performed by the code developer the level of support needed would be low to moderate, otherwise some form of training would be required.

[d] The computer code implementation of each methodology generates extensive results summaries that require post-processing.

[e] For example, refer to U.S. PSA Peer Review Guidelines (e.g. U.S. NRC Regulatory Guide RG 1.200), https://www.nrc.gov/docs/ML2023/ML20238B871.pdf

[f] Implicitly accounted for in OPEX databases. The DDM accounts for different material grades. PFM and I-PPoF explicitly account for the chemical and mechanical property data.

8. VALIDATING PROBABILISTIC OUTPUTS

8.1. BASIC CONSIDERATIONS

Invariably, piping reliability analyses produce frequency of failure being $<$ or $\ll 10^{-4}$ per piping component and reactor operating year (ROY), and sometimes with large uncertainties. Some form of validation against different modelling approaches is warranted to determine the level of realism that has been achieved. In order to establish a meaningful basis for comparing results obtained using different methodologies it becomes necessary to start by defining the specific figures of merit that are to be calculated. In the context of PSA and risk-informed applications, piping reliability typically is considered as part of the initiating event frequency modelling element. The PSA parameter of interest can be expressed in the form of a cumulative pipe break frequency versus a set of consequence threshold values expressed by an EBS with dimension of mass flow rate [kg/s] or a corresponding diameter of the hole in the pipe wall [mm] that would produce a certain flow rate. Therefore, the pipe failure mode of concern could involve a spectrum of consequences, from small through-wall leaks up to a major structural failure such as a double-ended guillotine break. Embedded in an analysis are multiple sets of input parameters that account for the effects of different degradation mechanisms on structural integrity, the effectiveness of ISI and leak detection, the effectiveness of degradation mitigation techniques, and the chemical and mechanical properties of the material.

The process for validating probabilistic outputs is a function of a selected technical approach that is used to calculate the reliability metrics, regulatory requirements, and applicable codes and standards. An underlying assumption in validation processes is that the implementation of calculation algorithms has been independently verified and documented. The validation step is concerned with how selected input parameters are justified, the sensitivity of results to the different assumptions that are used to establish the technical bases for input parameters and the treatment of uncertainties.

8.2. PROBABILISTIC ACCEPTANCE CRITERIA

Developing probabilistic acceptance criteria for pipe failure frequencies and conditional failure probabilities is a complex, lengthy process. It involves top-down/bottom-up considerations and the development of consensus guidelines on methods implementation. A top-down consideration could be a system level reliability target, and a bottom-up consideration could be component specific RIM processes that need to be applied in order to meet a system level reliability target. The processes for establishing acceptance criteria and the acceptability of methodology and its implementation are closely intertwined. Regulatory frameworks, and international and national codes and standards for fitness for service assessment, operability determination and PSA influence this development process. Embedded in it are requirements for what constitutes acceptable methodologies to be used in meeting the acceptance criteria and development of frameworks for how to conduct a peer review of piping reliability analyses [122−126].

Probabilistic acceptance criteria are used to guide or inform decisions involving, for example:

- Acceptability for continued operation for another fuel cycle, given a pre-existing, non-leaking pipe flaw;
- Acceptability of changing an existing ISI programme;
- Acceptability of a new RIM programme;
- Acceptability of the risk posed by a degraded piping component.

An objective of an acceptance criterion is to assess probabilistic results against it. For example, if it can be demonstrated that the assessed pipe failure frequency as represented by a probability density function is less than a certain value, then justification may exist for a favourable decision. Within the intent of an assessment, complications arise because of the uncertainties in the reliability metrics.

8.2.1. Defence in depth considerations

Acceptance criteria for probabilistic evaluations is to carefully consider the intent of the evaluation with respect to the concept of defence in depth (DiD). The five DiD levels that are discussed in the IAEA Specific Safety Requirements No. SSR-2/1 [127] are summarized in Table 7.

If the objective of a probabilistic evaluation is to demonstrate that an in-service failure of a component containing a detected flaw is unlikely, the acceptance criteria are to provide a level of confidence comparable with the margins established in the original design code. This would be an example of a DiD level 1 assessment that would be comparable to a traditional deterministic structural integrity evaluation of a flaw. The determination of an appropriate acceptance criterion for such applications is not a trivial task since nuclear design codes for pressure boundary components typically do not establish probabilistic safety margins. For this reason, the application of probabilistic evaluations for the evaluation of detected flaws is currently limited in the nuclear industry.

Deterministic safety margins in design codes establish limits on design loads and material resistance parameters to ensure a low likelihood of component failure. There is an inherent, and usually unquantified, probability of failure associated with those safety margins. The objective of the development of a probabilistic acceptance criterion for component structural integrity applications focuses on establishing a probabilistic criterion that would generate similar safety margins. This is illustrated in Fig. 31 [128].

TABLE 7. THE IAEA LEVELS OF DID

Level	Purpose	Requirements
1	Prevent deviations from normal operation and the failure of items important to safety	The plant be soundly and conservatively sited, designed, constructed, maintained and operated in accordance with quality management and appropriate and proven engineering practices
2	Detect and control deviations from normal operational states in order to prevent anticipated operational occurrences at the plant from escalating to accident conditions	The provision of specific systems and features in the design, the confirmation of their effectiveness through safety analysis, and the establishment of operating procedures to prevent such initiating events, or otherwise to minimize their consequences, and to return the plant to a safe state
3	Inherent and/or engineered safety features, safety systems and procedures be capable of preventing damage to the reactor core or preventing radioactive releases requiring off-site protective actions and returning the plant to a safe state	Engineered safety features and accident procedures
4	Mitigate the consequences of accidents that result from failure of the third level of defence in depth	Preventing the progression of such accidents and mitigating the consequences of a severe accidents
5	Mitigate the radiological consequences of radioactive releases that could potentially result from accidents.	The provision of adequately equipped emergency response facilities and emergency plans and emergency procedures for on-site and off-site emergency response

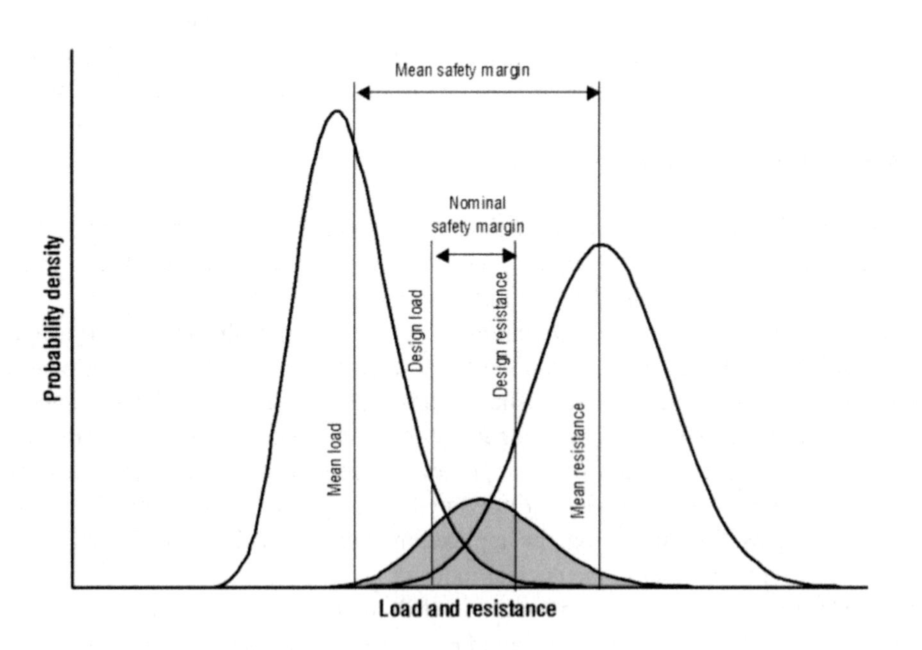

FIG. 31. Deterministic and probabilistic based design safety margins (adapted from [128]).

Guidance on the development of acceptance criteria for DiD level 1 may be obtained through an examination of civil structure design codes. For example, the American Society of Civil Engineers (ASCE) Standard, ASCE 7-16 [128], specifies target annual probability of failures for general structural design based on different reliability indices for different risk categories for 50 year service lives, excluding loads from extraordinary events such as earthquake or tsunami to excluding earthquake, tsunami, flood and loads from extraordinary events. These target annual probabilities of failure range from 1.25×10^{-4} per year for failures that do not occur suddenly and pose a low risk to human life, down to 1×10^{-7} per year for sudden failures that pose a high risk to human life. The probabilistic criteria in civil design codes normally limit both the annual probability of failure as well as the cumulative lifetime probability of failure. This is useful for systems or components subject to ageing. The annual failure probability limits the year over year change to ensure that the rate of degradation remains acceptable. The annual targets are calculated assuming a specified operating life (50 years in the case of ASCE 7-16).

The combination of the annual failure probability and the specified operating life limits the end-of-life cumulative probability of failure. Similarly, European Standard EN 1990:2002 [129] provides annual and 50 year probability of failure targets. The annual targets range from 1.3×10^{-5} to 1.0×10^{-7} and the 50 year targets range from 4.8×10^{-4} to 8.5×10^{-6} depending on the consequences of a failure.

The Electric Power Research Institute (EPRI) report MRP-362 [130] suggests an approach to quantify the annual probability of failure inherent in the deterministic design margins for an ASME Section III Class 1 pressure boundary component. The report was prepared to support the development of the ASME Section XI code case N-838, Flaw Tolerance Evaluation of Cast Austenitic Stainless Steel Piping [131]. For a given service level transient, an annual probability of failure of 1×10^{-6} was derived considering the probability of exceeding the material flow strength in piping containing the allowable flaw sizes in Table C-5310-1 of ASME Section XI. It is important to note that the code case defines target flaw sizes for the qualification of non-destructive examination tools and is not intended for use in flawed piping assessments. However, this is potentially a useful example to demonstrate how probabilistic acceptance criteria might be developed to maintain consistency with deterministic design requirements.

There are several examples of nuclear industry PFM evaluations using acceptance criteria derived or proposed based on the impact of a component failure on the core damage frequency (CDF) using DiD level 3 considerations.

The inter-dependence between pipe failure and CDF is described in Ref. [132]. When the output from a PFM code corresponds to an annual safety class 1 total pipe rupture frequency less than 1×10^{-6} per year the US NRC Regulatory Guide 1.174 criterion for acceptable increases in core damage frequency could be seen as satisfied [134]. Furthermore, the authors of [132] recognize the inherent uncertainties in piping reliability and that the pipe rupture frequency is represented by a probability density function and that "the acceptance criterion of 1×10^{-6} failures per year was developed with the understanding that it is to be assessed against the 95% confidence level of mean results. When an analysis yields results that are close to the acceptance criteria it may be necessary to further address uncertainty."

The US NRC has implemented a significance determination process to determine the risk significance of degraded or failed components. The methodology for assigning risk metrics to an operational event and inspection findings is presented in [133–135]. The significance determination process uses change in core damage frequency (ΔCDF) and change in large early release frequency (ΔLERF) as risk metrics. This process provides an initial screening to identify those regulatory inspection findings that do not result in a significant increase in plant risk and thus need not be analysed further (a 'green' finding). Remaining inspection findings, which may have an effect on plant risk, are subjected to a more thorough risk assessment. The final outcome of the review, evaluating whether the finding is green, white, yellow or red, is used to determine further regulatory actions that may be needed. The colours assigned to significance determination process findings are defined as follows:

— Red (or high safety significance) when change in risk (ΔCDF) is greater than 1×10^{-4};
— Yellow (or substantial safety significance) when change in risk is in the range of $1 \times 10^{-5} < \Delta CDF \leq 1 \times 10^{-4}$;

— White (or low to moderate safety significance) when change in risk is in the range of $1 \times 10^{-6} < \Delta\text{CDF} \leq 1 \times 10^{-5}$;
— Green (or very low safety significance) when $\Delta\text{CDF} < 10^{-6}$ or when $\Delta\text{CDF} \ll 10^{-6}$.

Calculating the change in risk given pipe failure can be done using the following equations:

$$\Delta\text{CDF} = \Sigma_i \Delta F_i \times \text{CCDP}_i \quad (1)$$

$$\Delta F_i = \Delta f \times \text{CFP}_i \quad (2)$$

$$\Delta f = f_{\text{New}} \times f_{\text{Old}} \quad (3)$$

The assumption is that a pipe failure (e.g. small leak) is a precursor to a LOCA of category *i* (small to very large), depending the magnitude of failure if a rapid flaw propagation occurs. These equations represent the change in CDF following a discovery of a flawed pipe. The failure could be a significant crack that was missed during ISI or it could be an active through-wall leak that necessitates a reactor shutdown. The definitions of the respective terms of the equations above are given in Table 8.

The Nuclear Structural Integrity Working Group has proposed back-calculating a target reliability of initiating events from the CDF as a possible method for the development of probabilistic acceptance criteria [136]. A detailed description of the concept is not provided in the Working Group document, but reference is made to a technical paper [137].

It is recognized that methods that use the CDF and the related risk metrics as the basis for deriving a probabilistic acceptance criterion for the assessment of a component may not provide a reliability target that is sufficient to meet the structural integrity objectives of the component design code (DiD level 1). Referring to Eq. (2), the contribution of the initiating event to the CDF is a function of the initiating event frequency and the CCDP. If the CCDP is high and the CDF contribution is limited to a low value, the resulting initiating event frequency will also be low. In such scenarios, it could be possible to generate a reliability target that is sufficient to ensure a very low probability of component failure during the intended operating life that would align with the intent of the design code. For example, for a PWR

TABLE 8. INPUT PARAMETERS TO CHANGE-IN-RISK CALCULATION

Parameter	Definition
ΔCDF	Change in CDF due to occurrence of degraded safety class 1 piping component. Could be a non-through-wall defect (e.g. circumferential crack) or minor leakage
ΔF_{i}	Change in frequency of category *i* LOCA initiating event frequency due to occurrence of degraded condition.
Δf	Change in pipe failure frequency due to the discovery of a new degraded condition
f_{New}	Plant specific failure frequency based on the state of knowledge immediately **after** discovery of a degraded condition
f_{Old}	Plant specific failure frequency based on the state of knowledge **before** discovery of a degraded condition
CFP_i	CFP that the new degraded condition led to a category *i* LOCA
CCDP_i	Conditional core damage probability given a category *i* LOCA

reactor vessel assessment, a rupture of the reactor vessel would likely generate a CCDP close to 1.0. Therefore, limiting the CDF contribution to a low value would result in a low allowable failure frequency. In contrast, the rupture of an individual pressure tube in a CANDU reactor core would result in a much lower CCDP because of the small diameter of the tube and mitigating safety systems. If the CCDP is low, a relatively high pressure tube failure frequency could still generate a low CDF contribution. While relatively high pressure tube failure frequencies may not impact plant safety as quantified by the effect on the CDF, it would not be desirable to operate a primary heat transport system with a high potential for primary heat transport system failures.

8.2.2 Definition of failure limit state

When establishing acceptance criteria from PFM applications it is important to ensure that the definition of failure is clearly defined, especially if results of different assessments will be combined or compared for risk-informed decision making. For example, consider the information from three different PFM applications summarized in Table 9. While all three applications propose the use of the same numerical value, 1×10^{-6}, the definitions are different in each case. There are three key observations arising from the comparison:

- The EPRI MRP-362 criteria [130] use an annual probability, while other examples in Table 16 use annual frequencies. As discussed in Section 8.2.3, annual probability and frequency are not equivalent reliability parameters by definition. These parameters are only approximately equivalent for special cases established using annual timescale and when the numerical values are small.
- The EPRI MRP-362 criteria apply to an evaluation of a single crack subject to a single service level load. For the NUREG-1874 [138] and future xLPR [139] applications, the acceptance criteria would be compared to a summation of frequencies for all individual flaws in the system, considering all loading scenarios.
- The definition of a failure for the NUREG-1874 criteria is the propagation of a crack through the wall thickness. The evaluation does not consider the stability of the through-wall cracking. The criteria for the EPRI MRP-362 and proposed xLPR applications allow for the consideration of through-wall cracking stability.

As a result of the differences associated with these three observations, the acceptance limits for these three examples would generate different reliability targets.

8.2.3 Selecting an appropriate reliability parameter

When defining acceptance criteria for probabilistic assessments, consideration should be given to the selection of the appropriate reliability parameter. The information below is not intended to provide a comprehensive review of reliability theory but illustrates some of the concepts that should be considered when developing acceptance criteria.

The examples discussed previously used annual failure probabilities, failure frequencies and cumulative lifetime probabilities. Reliability assessments are generally classified into one of two categories, assessments for non-repairable or repairable systems (a system may be an individual component or combination of components depending on the problem definition) [139]. Non-repairable systems are those which are concerned with first failure. After a system failure is observed an end state is reached and it is impossible to return to service. In such scenarios, mission reliability or the annual probability of failure over a specified interval are appropriate reliability metrics [140]. Failure rate or failure frequency are reliability metrics that apply to repairable systems. After equipment failure is experienced, a corrective action can be implemented (i.e. repair or replacement) so the equipment can be returned to operation.

For non-repairable systems, the mission is the defined evaluation interval for the assessment. If the mission is defined as the lifetime of the system from the start of operation at time $t = 0$ to a specified end-of-life date then the mission probability of failure equates to the cumulative lifetime probability of failure. For a cumulative failure distribution function $F_t(t)$, the mission probability of failure for any interval (t_1, t_2) can be denoted $P_f(t_1, t_2)$. In the non-repairable problem, definition $P_f(t_1, t_2)$ is to be conditioned to account for the non-zero probability of failure up to time t_1 using:

$$P_f(t_1,t_2)=\frac{F_t(t_2)-F_t(t_1)}{1-F_t(t_1)} \tag{4}$$

The mission reliability, $R_t(t_1, t_2)$ is calculated using:

$$R_t(t_1,t_2)=1-P_f(t_1,t_2)=\frac{1-F_t(t_2)}{1-F_t(t_1)} \tag{5}$$

If the mission time is 1 year, Eqs (4) and (5), the annual probability of failure and the annual reliability is calculated.

The potential significance of the conditioning effect for a hypothetical probabilistic assessment for a non-repairable system which generates the cumulative distribution function is shown in Fig. 32. Using Eq. (4), the mission probability of failure calculations is summarized in Table 10 for the 5 to 10 year interval and the 25 to 30 year interval. The mission probability of failure accounts for the higher likelihood of a failure occurring between years 25 and 30 given it has survived up to year 25.

When $F_t(t_1)$ is very low, then $P_f(t_1, t_2) \approx F_t(t_2) - F_t(t_1)$, but the mission probability of failure can be significantly underestimated and the mission reliability significantly overestimated if that is not the case. In many nuclear industry PFM applications, the objective is to demonstrate that probability of failure remains very low over the expected operating life of the component, so $F_t(t_1)$ is often very small. As a result, the conditioning effect in Eqs (6) and (7) may be insignificant, but this is not to be assumed without confirmation for a specific application.

TABLE 9. COMPARISON OF ACCEPTANCE CRITERIA PROPOSED FOR DIFFERENT PFM APPLICATIONS

Proposed application	Intent of application	Numerical acceptance criteria	Definition of acceptance limit
EPRI MRP-362	Evaluate design intent probability of rupture for ASME class 1 cast austenitic stainless steels piping	1.0×10^{-6}	Annual probability of rupture per crack for a service level load
NUREG-1874	Demonstrate probability of failure of a class 1 reactor pressure vessel due to pressurized thermal shock is extremely low	1.0×10^{-6}	Annual frequency of occurrence of through-wall cracking cumulative for all welds in the vessel
xLPR	Demonstrate probability of failure of class 1 piping with dissimilar metal welds is extremely low	1.0×10^{-6}	Annual frequency of rupture cumulative for all susceptible welds in an NPP

For repairable systems, the failure rate or failure frequency is defined as the expected number of failures per unit time. At time t, the failure rate, $\eta(t)$, can be calculated based on [141–143], for a probability density function for the mean time between failures, X, represented by $f_x(t)$, as follows:

$$\eta(t)=f_X(T)+\int_0^t \eta(t-x)f_X(x)\mathrm{d}x \tag{6}$$

The homogenous Poisson process model is widely used to model recurring events [137]. Parameter X is an exponential random variable and the distribution of the number of failures, $N(t)$, follows the Poisson distribution, where:

$$P\left[N(t)=k\right]=\frac{(\lambda t)^k e^{-\lambda t}}{k!} \tag{7}$$

The expected number of failures is a function of the failure rate and time, $E[N(t)] = \lambda t$ and the failure rate is a constant value, $\eta(t) = \lambda$. Because the failure rate is constant, the homogenous Poisson process model is not appropriate for modelling equipment subject to ageing related degradation where the failure rate is expected to increase with time. However, for highly reliable equipment, λ tends to be small and the probability of failure in time, t, can be approximated as $P_f(t) \approx \lambda t$ leading to the often used assumption that the annual probability of failure $P_f(t = 1) \approx E[N(t = 1)]$. The validity of this assumption has to be verified for a specific application if it is adopted as the basic for acceptance criteria for probabilistic assessments for repairable equipment. If ageing effects are significant failure, other probability models which permit

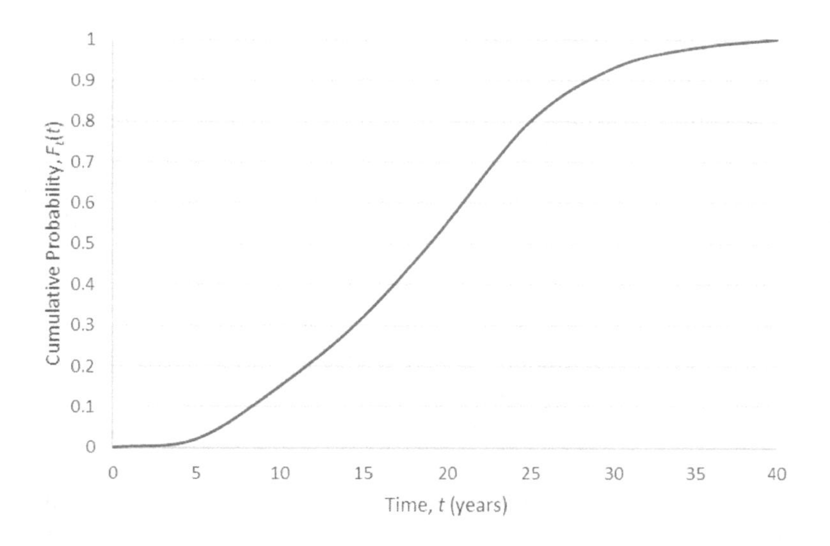

FIG. 32. Cumulative distribution function for a hypothetical probabilistic assessment for a non-repairable component.

TABLE 10. MISSION PROBABILITY OF FAILURE CALCULATIONS FOR THE EXAMPLE DEPICTED IN FIG. 32.

t_1	t_2	$F_t(t_1)$	$F_t(t_2)$	$F_t(t_2) - F_t(t_1)$	$P_f(t_1, t_2)$
5	10	0.020	0.150	0.130	0.132
25	30	0.800	0.930	0.130	0.650

non-constant failure rates, such as the Weibull model, are considered. The failure rate, $\lambda(t)$, for a Weibull model can be calculated as follows:

$$\lambda(t)=\frac{\beta}{\upsilon}\left(\frac{t-\gamma}{\upsilon}\right)^{\beta-1} \tag{8}$$

where β is the shape parameter, υ is the scale parameter and γ is the location parameter for the distribution. For $\beta < 1$, the failure rate would decrease with time, for $\beta = 1$, the failure rate would be constant with time and for $\beta > 1$, the failure rate would increase with time.

8.2.4. Summary

Development of acceptance criteria for probabilistic evaluations of the risk associated with the failure of NPP components is not a trivial task. When acceptance criteria are based on the effect that a component failure has on the CDF, this may not provide sufficient operational reliability targets for all pressure boundary components. Inconsistent definitions of the failure limit state can present challenges in comparing the risk associated with different component failures or combining component failure probabilities for a system. The selection of the appropriate reliability target needs to account for the nature of the modelled system. Advancement of the application of PFM in the nuclear industry would benefit from a strategic and systematic review of appropriate acceptance criteria and adoption of the criteria in industry codes and standards.

8.3. SENSITIVITY ANALYSIS AND INFLUENCE PARAMETER RANKINGS

Sensitivity analysis and importance ranking can be utilized to help improve the validity of piping reliability models. These types of evaluations are performed in order to: (a) analyse the impact of each input parameter and modelling assumption on the magnitude of the pipe reliability estimate (reliability importance); and (b) analyse the contribution of each input parameter and modelling assumption to the total epistemic uncertainty of the pipe reliability estimate (uncertainty importance). The reliability importance can demonstrate how the pipe reliability model responds to a perturbance of individual input parameters and modelling assumptions; hence, it can help the analyst better understand the model behaviour and detect an abnormal response that may indicate an error in input data or models. The uncertainty importance can suggest, if the validity of the pipe reliability estimation needs to be improved, which input data and/or modelling assumptions are prioritized for further refinements.

In the DDM approach, sensitivity analyses are performed at different levels of an analysis. A sample based uncertainty method is used to assess variations and uncertainties for each key parameter. In preparing the input parameter sets, different assumptions about the prior state of knowledge (i.e. prior failure rate distributions) are tested in order to assess the effect of a chosen prior on the posterior failure rate distribution. In the analysis of the effect of RIM on the probabilistic failure metrics, sensitivity studies are performed by varying the value of the probability of detection and leak detection intervals. In developing a CFP model, the parameters of the beta distribution may be obtained using different technical approaches, including expert elicitation and PFM.

PFM simulates the behaviour of cracked structures and transfers uncertainties from input parameters to a failure frequency, expressed in terms of the probability density function. The assessment of the rupture probability of highly reliable piping can be obtained. For the appropriate probabilistic modelling of a structure including the uncertainties, but also for the analysis of PFM application cases, the question is, what input parameter has higher impact on computed failure probability, and which have minor impacts. This question is related to sensitivity measures or importance factors of the input parameters and their ranking based on their influence [144, 145]. The following are the methods for the ranking of relevant influence parameters [24, 146]:

- *Amplification ratio*. The amplification is useful for identifying basic variables that have a strong effect on the resulting estimate of failure probability.
- *Direction cosine method*. This method is based on the position of the most probable failure point, and it provides a measure of the sensitivity or importance of the failure probability to the corresponding random variable.
- *Degree of separation*. The mean value (μ_i) and the standard deviation (σ_i) are estimated from all realizations sampled in the simulation rather than those of user-defined input distributions.
- *Separation of uncertainty method*. A two loop PFM code architecture for the treatment of uncertainties allows for separate treatment of the aleatory and epistemic uncertainties.
- *Sample based uncertainty method*. At least three levels are considered for each parameter based on the anticipated variations and uncertainties: low/baseline/high, where baseline represents the original parameter set; low numerically smaller; and high numerically higher values — alternatively, the low/high nomenclature can refer to the expected structural performance or risk.

The I-PPoF framework conducts sensitivity analyses with respect to both reliability and uncertainty importance of input parameters. For the reliability importance, local sensitivity methods can be used since their interpretations with respect to the model's behaviour are straightforward and can be visualized. For the uncertainty importance, I-PPoF utilizes the cumulative distribution function based moment-independent global method due to its capability of accounting for uncertainties associated with the model input and outputs as well as the non-linearity and interactions among input parameters inside the model. For instance, Beal et al. [147] conducted local sensitivity analyses using the tornado diagram and spider plot techniques to study the reliability importance of the RIM input parameters and a global sensitivity analysis to identify the uncertainty importance of those input parameters. Beal et al included a global sensitivity analysis for uncertainty importance as one of the steps in the probabilistic validation. In their algorithm, if the epistemic uncertainty for the pipe reliability estimation does not satisfy the probabilistic acceptance criteria, global sensitivity analysis is conducted to identify input data and modelling assumptions that are the dominant sources of epistemic uncertainties and can be refined to reduce the overall uncertainty.

8.4. EXPERIMENTAL WORK AND MODEL VALIDATION

An objective of experimental work is to solve knowledge base gaps in materials science and fracture mechanics. The experimental work includes laboratory tests, computer simulations (e.g. acoustic-structural simulation, coupled computational fluid dynamics and structural simulation), and large scale tests of mock-ups subjected to different environments (pressure, temperature, chemistry parameters) and loading conditions. The results from the experimental work can be essential in the characterization of uncertainties in input parameters. This becomes especially important when analysing situations for which there is no OPEX data available on which to base the input parameter selections. This subsection has a narrow scope. It highlights some of the outcomes from the extensive research to determine the susceptibility of certain nickel base alloys to SCC in a high temperature environment; specifically, the performance of Alloy 690 and its weld metals Alloy 52 and 152 in a PWR primary system operating environment. Selected chemical and mechanical properties of nickel base materials are summarized in Table 11.

In a PWR primary system water environment, intergranular SCC in wrought nickel base materials is commonly referred to as primary water SCC (PWSCC). The predominant nickel base alloys in WCRs are Alloy 600 and Alloy 690. The former was developed in the 1950s and the latter in the 1960s, with full scale production beginning in the 1960s and 1970s, respectively. The occurrence of PWSCC in high purity water has been extensively studied since the first reported observation in laboratory tests performed

TABLE 11. SELECTED CHEMICAL AND MECHANICAL PROPERTIES OF ALLOYS 600 AND 690

Mechanical properties									
Temperature		Alloy 690				Alloy 600			
		Yield strength		Tensile strength		Yield strength		Tensile strength	
°F	°C	ksi	MPa	ksi	MPa	ksi	MPa	ksi	MPa
1000	565	45.5	314	105.5	727	28.5	196	84.0	579
1200	650	46.1	318	108.5	748	26.5	183	65.0	448
1400	760	46.5	321	103.5	714	17.0	117	27.5	190
Chemical properties									
Nickel		≥58%				≥72%			
Chromium		27.0–31.0%				14–17%			
Iron		7.0–11.0%				6.0–10.0%			
Carbon		0.05% max.				≤0.15%			

at the Commissariat à l'Énergie Atomique in 1959 [148, 149]. These tests were performed on Alloy 600 material. Alloy 600 nevertheless became a preferred alloy when the WCRs were designed and constructed in the 1960−1990 time frame. The first Alloy 600 PWSCC events in operating plants were reported in the early 1980s.

The OPEX data on PWSCC is extensive. There have been on the order of 200 Alloy 600 failures involving significant part through-wall cracking or through-wall cracking of piping components, and an even larger number of PWSCC failures of non-piping passive components. Excluding steam generators, for piping and non-piping passive components, the overall failure population exceeds 500 events. One of the insights that has been obtained from this experience is that some of the PWSCC cases occurred after relatively long incubation times (>100 000 equivalent full power hours). Alloy 690 and its weld metals (Alloy 52 and 152) was developed to resist SCC, and it has almost universally been the preferred replacement for Alloy 600/82/182. As of this writing, there are no known failures after several decades of field experience with Alloy 690. Without OPEX data, the question is, how can the resistance of Alloy 690 to primary water SCC be characterized? Crack initiation data are obtained from laboratory tests that are based, for example, on the constant extension rate tensile technique [150]. The applied strain rates depend on the test objectives and other experimental variables. There are different technical approaches to determine the factor of improvement for this alloy relative to Alloy 600:

— Weibull analysis of Alloy 600 cumulative fraction of cracking by end of a test period, and Weibayes (or Bayesian Weibull) analysis of Alloy 690 specimens that did not develop cracking by end of a test period. In the Weibayes analysis, assumptions are made about the slope parameter based on judgement, for example, the slope parameter for Alloy 690 may be assumed to be the same as for Alloy 600. This technique is explained in [150]; it assumes that ample test data are available to support the statistical analysis.

— With insufficient test data, the factor of improvement is calculated as the fraction of the Alloy 690 test time to the time of first cracking in an Alloy 600 specimen [150].

8.4.1. Alloy 690 research programmes (years 1985–2020)

In a laboratory setting, experimental tests are intended to replicate the environment that the materials are exposed to in an operating reactor. Because the incubation period of an environmental degradation mechanism may last for several years under typical operating conditions, laboratory tests are accelerated. Two techniques are generally used: an increased temperature and/or an increase in the applied stress. Standardized test protocols have been developed in order to ensure that test results performed by different laboratories are comparable. Reference [148] identifies more than 70 test protocols. Alloy 690 test programmes have been ongoing since the mid-1980s [151–155]. The many international test programmes have involved hundreds of tests using different standards and different types of specimens. Because of the multitude of test variables, the analysis and interpretation of results of these programmes is a complex undertaking.

A methodology for evaluating the test data has been developed by the Material Research Program (MRP) of the EPRI. Organized by the MRP, the results of an international primary water SCC Crack Growth Rate Expert Panel were published in 2018 [155]. The expert panel compiled a database of over five hundred Alloy 690 crack growth rate data points and over 130 Alloy 52/152 CGR data points from seven research laboratories These data points were evaluated and scored for data quality and assessed to determine the effects of parameters such as temperature, crack tip stress intensity factor yield strength and crack orientation. According to this expert panel, the suggested conservative factors of improvement are 38 for Alloy 690 versus Alloy 600, and 324 for Alloys 52/152 versus Alloys 82/182. Figure 33 presents a summary of two of the data sets evaluated by the expert panel; one data set with a 600 MPa yield strength and another with a 12% cold worked limit. In a subsequent revision, the suggested conservative factors of improvement were given as 48 for Alloy 690 and 231 for Alloy 52/152 [155].

8.4.2. Estimates of Alloy 690 versus Alloy 600 factor of improvement

Covering multiple decades of experimental work, extensive published information on the Alloy 690 to Alloy 600 factor of improvement determination exists. The published work covers the different methods for factor of improvement determination, a wide array of test methods, test specimens and test variables. Active research remains ongoing and is conditioned by: (a) as of yet (year 2021) no evidence of Alloy 690/52/152 failures in the field, and (b) an increasing number of NPPs entering into periods of long term operation (>40 years). A factor of improvement is not arbitrarily selected. Selected values are to acknowledge the current state of knowledge, including the uncertainties in factor of improvement determination. In evaluating factor of improvement estimates the analyst is to differentiate thick-walled specimen data from thin-walled specimen data, as well as Alloy 690 from Alloy 52/152 specimen data.

9. INTERPRETING RESULTS

9.1. RESULTS PRESENTATION

The format for how to document results depends on the objectives of an analysis, including the relevant codes and standards requirements. As a minimum, a tabular summary of the results is to be presented that includes a characterization of the uncertainty in a calculated piping reliability parameter (i.e. mean, median, 5th percentile, 95th percentile and range factor which is defined as $(95th/5th)^{0.5}$). Graphical summaries can be in the form of the cumulative failure rate (or frequency) as a function of pipe

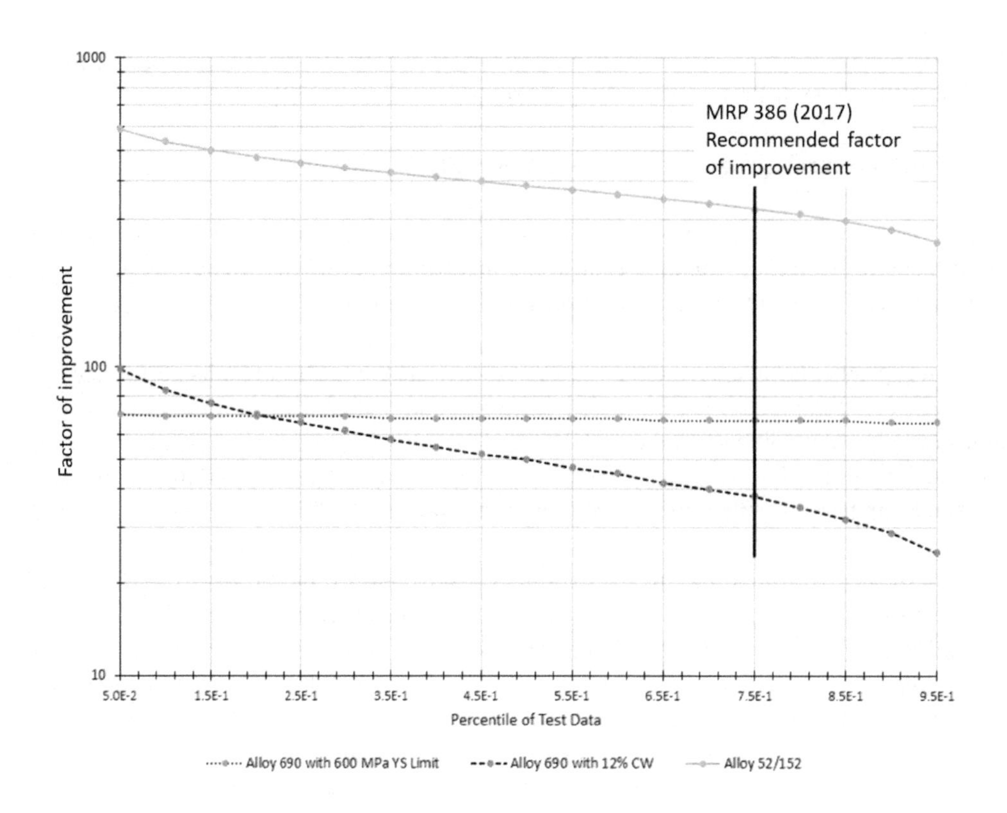

FIG. 33. Alloy 690 vs. Alloy 600 factor of improvement according to MRP-386.

break size (Fig. 34) or through-wall flow rates, or it can be in the form of a hazard function (e.g. annual pipe failure frequency versus ROY, Figs. 35 and 36). A hazard function can be viewed as a probability density function divided by the survival function; or the instantaneous frequency of failure within a narrow time frame.

Presenting results in terms of frequency versus break sizes implies two different underlying assumptions. A first assumption is a double-ended guillotine break only type of failure, meaning that if through-wall cracking develops the system pressure would cause the crack to propagate into a full double-ended guillotine break. Under the first assumption the higher failure frequency is attributed to a greater likelihood of double-ended guillotine breaks on small diameter pipes. A second assumption is referred to as the continuum failure model in which an EBS of any size up to and including a double-ended guillotine break can occur on any pipe. In this model, the higher frequency associated with small breaks is attributed to the combination of double-ended guillotine breaks on small diameter pipes and small EBS on large pipes.

9.2. RISK CHARACTERIZATION OF PIPE FAILURES

As an introduction to Sections 9.3 and 9.4, this section discusses a pipe failure event analysis to illustrate how a risk characterization of a failed pipe can be performed and how the reliability metrics relate to and challenge probabilistic acceptance criteria. The intent of a risk characterization can be to quantify the change in CDF due to a pipe failure that is discovered during routine plant operation. The basic elements of a risk characterization process are illustrated in Fig. 37.

Probabilistic failure metrics for risk characterizations include the CCDP and change in core damage frequency, ΔCDF [155]. Different national codes and standards for risk characterization exist. In this example, the CCDP is used as a measure of the core damage frequency at the time of the occurrence of a pressure boundary failure (PBF), for example, if a pipe failure were to produce a loss of coolant accident:

$$\text{CDF|PBF} = f_{\text{PBF}} \times \text{CCDP|PBF} \tag{9}$$

$$\Delta\text{CDF} = \Delta f_{\text{PBF}} \times \text{CCDP|PBF} \tag{10}$$

$$\Delta f_{\text{PBF}} = f_{\text{PBF-After}} - f_{\text{PBF-Before}} \tag{11}$$

9.2.1. Risk significance of pipe failures

The risk characterization process is exemplified by using a fictitious high cycle fatigue failure of a high pressure safety injection line connected to the primary pressure boundary piping in a PWR plant. The evaluation boundary of interest is shown in Fig. 38. The event is assumed to have occurred during normal plant operation, resulting in a controlled plant shutdown. A risk significance evaluation was performed after the fact.

The inside diameter of the pipe that leaked was 54 mm. The PSA model of the plant in which the pipe failure occurred classified a safety class 1 pipe failure having an EBS between 10 mm and 35 mm as a small LOCA, and EBS between 35 mm and 110 mm as a medium LOCA. Had the affected high pressure safety injection line suffered a double-ended guillotine type of break the effective break size would be about 75 mm. [157] Therefore, the high cycle fatigue failure was characterized as a precursor to a small or medium LOCA. The risk significance of the high cycle fatigue failure would be that it altered the plant specific state of knowledge about the frequency of a small LOCA or medium LOCA initiating event.

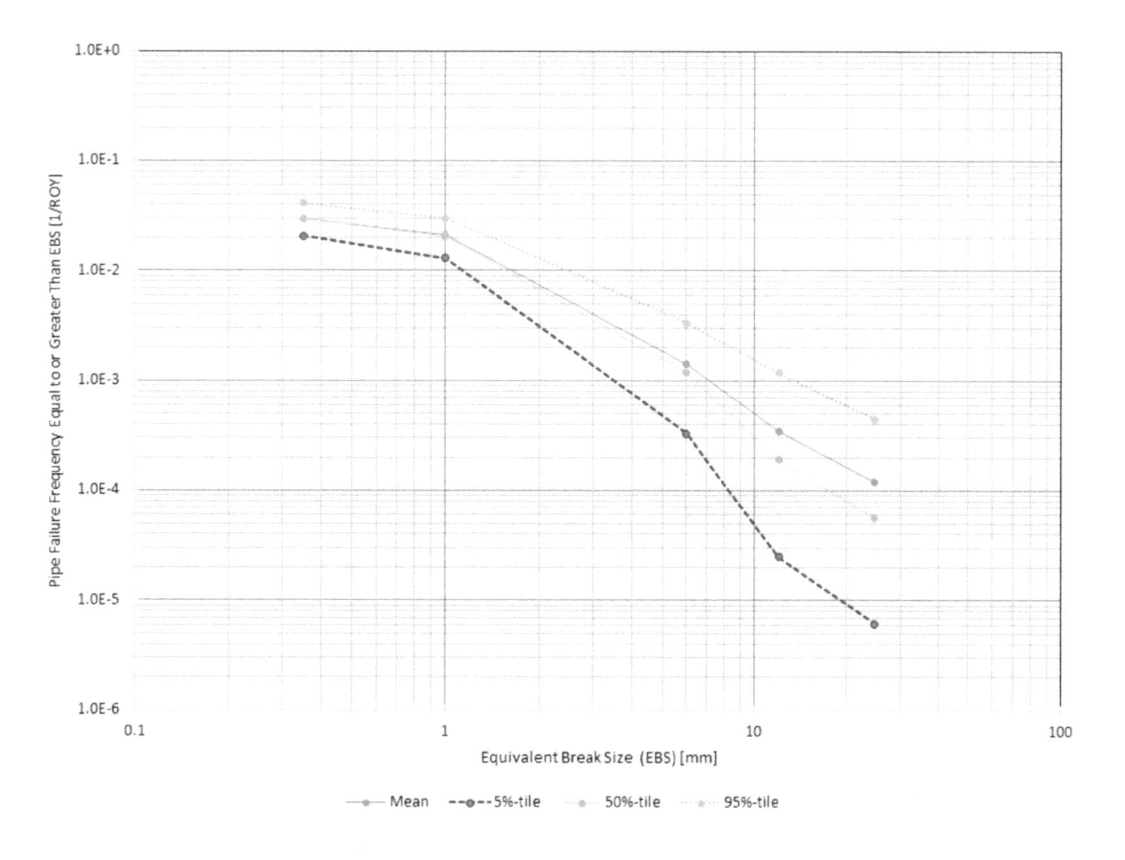

FIG. 34. System-level small diameter piping reliability results presentation.

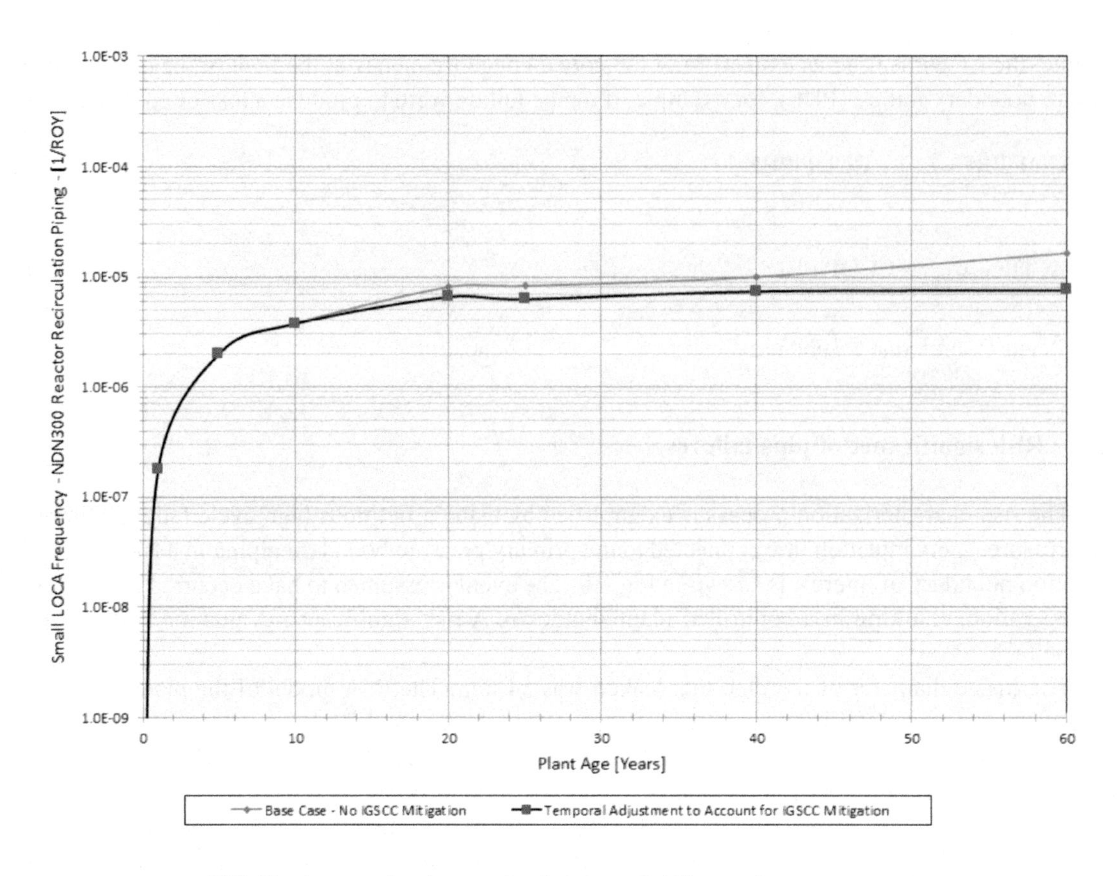

FIG. 35. An example of system-level piping reliability results presentation.

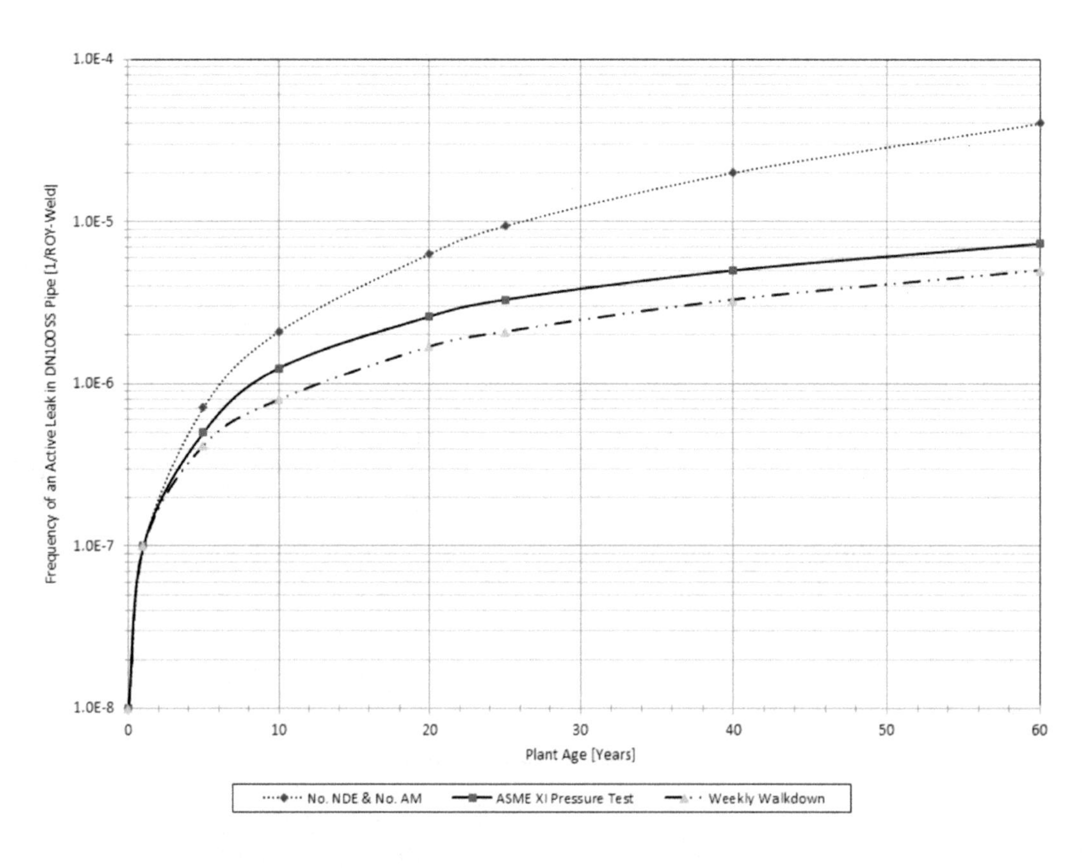

FIG. 36. An example of piping component level results presentation.

In this example, the approach to characterizing the risk significance of the leak event is to express the risk significance in terms of the change in CDF due to the leak event. A DDM is applied to calculate the relevant probabilistic failure metrics for the evaluation boundary (i.e. the pipe failure rate before and after the incident and the corresponding conditional failure probability).

Pipe failure rate estimates derived from OPEX are shown Table 12. By comparing the plant specific estimates before and after the leak event occurred, it is seen that the impact of the leak event is an increase in the high pressure safety injection line failure frequency of about 15%. Neither of the 26 high pressure safety injection line failures nor in any of the PWR pipe failures in the OPEX database, were the failures more severe than a relatively large leak (≤0.8 kg/s). In order to produce a small LOCA as defined in plant specific PSA models the failure would need to have an EBS of about 10 mm or larger (i.e. ≥7 kg/s). To produce a medium LOCA a break would have to exceed 110 mm in EBS (i.e. ≥100 kg/s).

In this example, the results of the LOCA Frequency Expert Elicitation Study (NUREG-1829) are used as the basis for the CFP [23]. This reference includes tabulations of LOCA frequencies for different systems and break sizes. In this example, a reverse engineering approach is applied as follows:

— Obtain the LOCA frequency distribution from NUREG-1829;
— Using the DDM, calculate a corresponding pipe failure rate;
— Derive a CFP as $\frac{\text{LOCA Frequency}}{\text{Pipe Failure Rate}}$.

A target LOCA frequency distribution is obtained by combining the geometric mean of the expert elicitation results[8] with the results of a DDM derived LOCA frequency distribution from Appendix D of NUREG-1829 [23]. This then becomes the mixture distribution as shown in Table 13. Next, the target LOCA frequencies are converted into CFP distributions using the failure rate distributions from the DDM derived results and after fitting the LOCA frequency distributions and failure rate distributions to lognormal distributions. Using this approach, the following relations are established:

$$\text{median}_{CFP} = \frac{\text{median}_{TLF}}{\text{median}_{FR}} \tag{12}$$

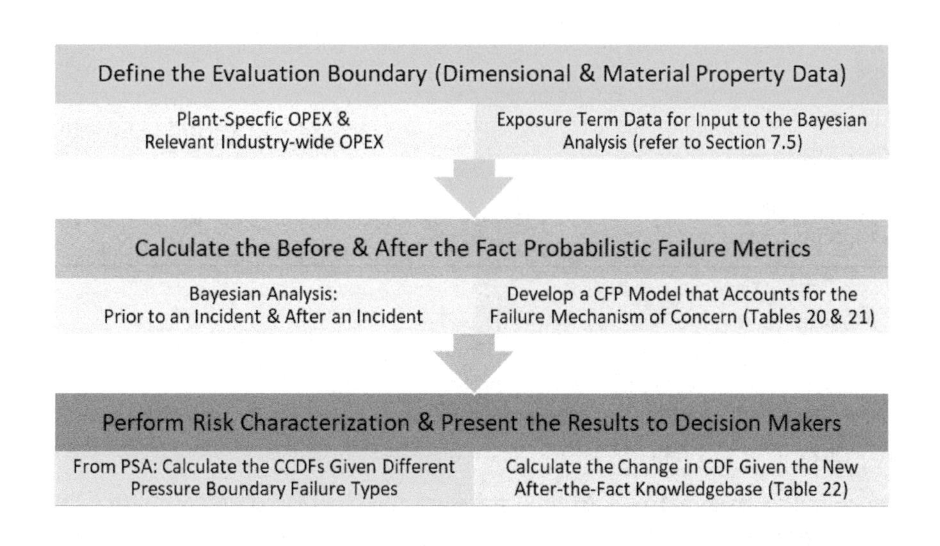

FIG. 37. Risk characterization process.

[8] Nine expert panel members produced nine different sets of LOCA frequency estimates that were obtained using different methods, including a DDM and PFM. The results are available from https://adams.nrc.gov/wba/ (ML080560011).

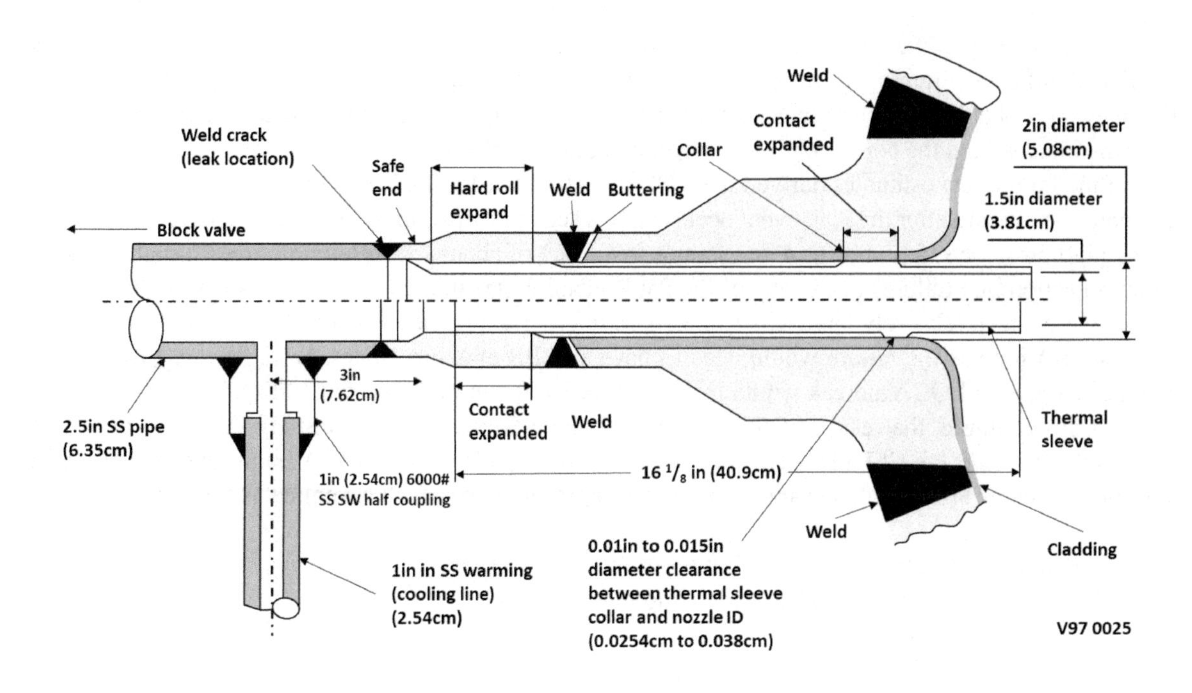

FIG. 38. Location of through-wall weld flaw (adapted from [156]).

TABLE 12. ESTIMATES OF HIGH PRESSURE SAFETY INJECTION LINE FAILURE RATE

HPI failure frequency case	OPEX failure count	Exposure [ROY]	Gamma distribution parameters				
			α	β	Mean	5%tile	95%tile
All PWR data pooled	26	6342	0.5	122	4.1×10^{-3}	1.7×10^{-5}	1.6×10^{-2}
PWR data excluding the plant where the failure occurred	19	6220	0.5	164	3.1×10^{-3}	1.3×10^{-5}	1.2×10^{-2}
λ_{Old} — Bayes' update before the leak event — plant specific	6	120.1	6.5	284	2.3×10^{-2}	1.0×10^{-2}	3.9×10^{-2}
λ_{New} — Bayes' update after the leak event — plant specific	7	121.6	7.5	285	2.6×10^{-2}	1.3×10^{-2}	4.4×10^{-2}

$$RF_{CFP} = e^{1.645\sigma_{CFP}} \tag{13}$$

$$\sigma_{CFP} = \sqrt{\left(\frac{lnRF_{TLF}}{1.645}\right)^2 - \left(\frac{lnRF_{FR}}{1.645}\right)^2} \tag{14}$$

The evidence in this example was 26 high pressure safety injection pipe failures involving cracks and leaks and 0 failures involving LOCAs. The Bayes' update was performed using priors from Table 14 and the binomial distribution for the likelihood function with the evidence of 0 ruptures out of 26 pipe failures. The resulting posterior CFP distributions are given in Table 15.

TABLE 13. HIGH PRESSURE INJECTION LINE LOCA FREQUENCIES FROM NUREG-1829

Distribution	LOCA category (NUREG-1829)	Break size [mm]	Mean	5%-tile	50%-tile	95%-tile
Geometric mean of NUREG-1829 [23]	1	≥15	1.3×10^{-5}	6.4×10^{-7}	5.5×10^{-6}	4.7×10^{-5}
	2	≥40	4.6×10^{-6}	1.5×10^{-7}	1.6×10^{-6}	1.7×10^{-5}
	3	≥75	7.2×10^{-7}	1.5×10^{-8}	2.1×10^{-7}	2.8×10^{-6}
Base case results Appendix D of NUREG-1829	1	≥15	1.6×10^{-5}	2.6×10^{-7}	3.9×10^{-6}	6.1×10^{-5}
	2	≥40	2.3×10^{-6}	3.3×10^{-8}	5.4×10^{-7}	9.0×10^{-6}
	3	≥75	9.2×10^{-7}	1.3×10^{-8}	2.1×10^{-7}	3.6×10^{-6}
Mixture distribution of experts and NUREG-1829 Appendix D results	1	≥15	1.4×10^{-5}	3.9×10^{-7}	4.7×10^{-6}	5.3×10^{-5}
	2	≥40	3.5×10^{-6}	5.5×10^{-8}	9.8×10^{-7}	1.4×10^{-5}
	3	≥5	8.1×10^{-7}	1.4×10^{-8}	2.1×10^{-7}	3.1×10^{-6}

TABLE 14. CUMULATIVE CFP DISTRIBUTIONS FOR HPI PIPE

Distribution	LOCA category	EBS [mm]	CFP distribution parameters			
			Mean	5%-tile	Median	95%-tile
Prior distribution for CFP	1	≥15	1.2×10^{-2}	5.8×10^{-3}	1.0×10^{-2}	1.8×10^{-2}
	2	≥40	3.0×10^{-3}	5.3×10^{-4}	2.1×10^{-3}	8.4×10^{-3}
	3	≥75	6.5×10^{-4}	1.1×10^{-4}	4.5×10^{-4}	1.8×10^{-3}
Posterior distribution using 0 ruptures in 26 pipe failures	1	≥12	1.1×10^{-2}	5.7×10^{-3}	9.9×10^{-3}	1.7×10^{-2}
	2	≥40	2.8×10^{-3}	5.1×10^{-4}	2.0×10^{-3}	7.6×10^{-3}
	3	≥75	6.3×10^{-4}	1.1×10^{-4}	4.5×10^{-4}	1.8×10^{-3}

TABLE 15. CHANGE IN PLANT SPECIFIC LOCA FREQUENCIES

Break size [mm]	LOCA frequencies prior to leak event		LOCA frequencies after leak event	
	Mean	95%-tile	Mean	95%-tile
15	2.4×10^{-4}	4.9×10^{-4}	2.8×10^{-4}	5.5×10^{-4}
40	6.4×10^{-5}	1.9×10^{-4}	7.3×10^{-5}	2.2×10^{-4}
75	1.5×10^{-5}	9.9×10^{-5}	1.7×10^{-5}	5.0×10^{-5}

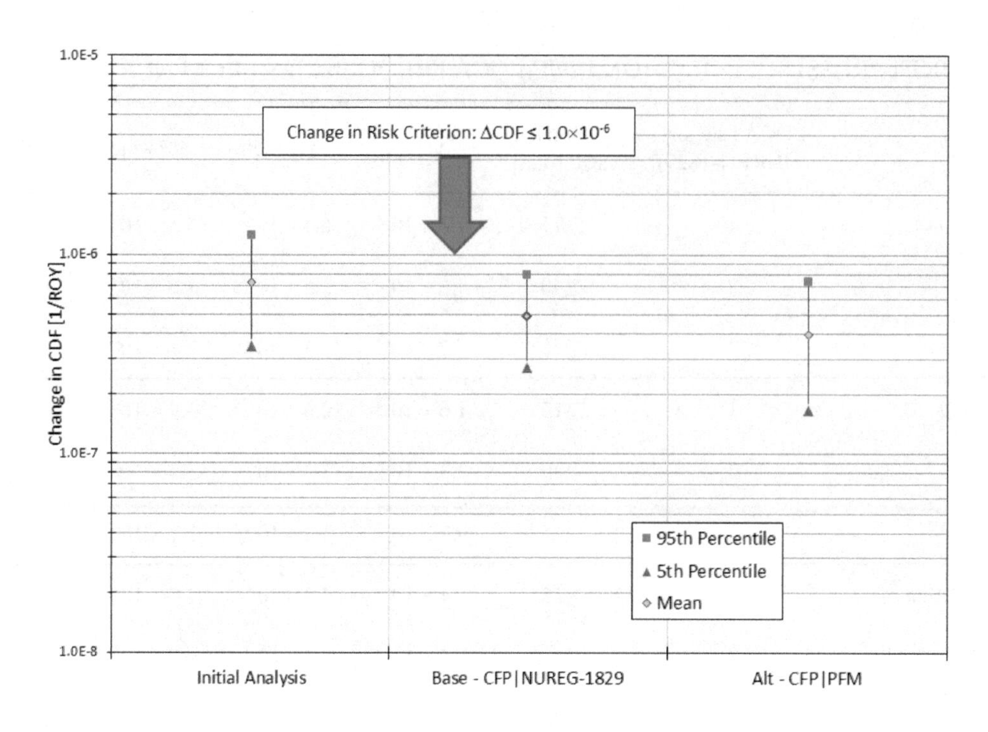

FIG. 39. Change in CDF due to a leak event.

9.2.2 Acceptability of risk characterization results

The results for the change in CDF due to changes in the state of knowledge before and after the occurrence of the fictitious pipe failure are shown in Fig. 39. Indicated in this figure are the results of three different calculation cases that were defined using different CFP assumptions.

The base case mean estimate of the change in CDF was estimated to be about 5×10^{-7} per year. It was based on the target LOCA frequency approach. An alternative calculation case used PFM calculations to generate CFP prior distributions. For illustrative purposes only, a simplified approach used an arbitrarily assumed CFP so that the upper bound estimate would be in excess of the risk criterion. This leads to the questions, what is the most appropriate way of justifying results and what are the underlying assumptions.

9.3. SENSITIVITY ANALYSIS

Sensitivity analysis is the study of how uncertainty in the output of a model can be apportioned to different sources of uncertainty in the model input [114–120, 146]. Using the example in Section 9.2.1, the model output parameters are influenced by assumptions about the prior pipe failure rate distribution (minor effect in the example) and the CFP model assumptions. The example based the CFP model on expert elicitation input and PFM.

9.4. UNCERTAINTY ANALYSIS

Uncertainty is defined as the representation of the confidence in the state of knowledge about the parameter values and models used in representing piping reliability. Uncertainty analysis focuses on identifying and quantifying the sources of uncertainty in the results of an analysis. The sources of uncertainty that are relevant to structural reliability analysis can be classified into two categories: (1) aleatory uncertainties, being associated with physical uncertainty or randomness; and (2) epistemic uncertainties, being associated with understanding or knowledge. Aleatory uncertainty refers to the natural randomness associated with an uncertain quantity and is often termed type I uncertainty in reliability analysis. Aleatory uncertainty is quantified through the collection and analysis of data. The observed data may be fitted by theoretical distributions, and the probabilistic modelling may be interpreted in the relative frequency sense. Epistemic uncertainty reflects the lack of knowledge or information about a quantity and is often termed type II uncertainty in reliability analysis. Key aspects of uncertainty treatment in structural reliability analysis are summarized below.

9.4.1. Aleatory vs. epistemic uncertainty

Aleatory uncertainty is attributed to inherent randomness, natural variation, or chance outcomes in the physical world; in principle, this uncertainty is irreducible because it is assumed to be a property of nature. Aleatory uncertainty is sometimes called random or stochastic variability.

Epistemic uncertainty is attributed to the lack of knowledge about events and processes. In principle, this uncertainty is reducible because it is a function of information. Epistemic uncertainty is sometimes called subjective or internal uncertainty, and divides into two major subcategories: model uncertainty and parameter uncertainty. When combined with the aleatory uncertainty, the epistemic uncertainty parameters provide a basis for quantitatively estimating the uncertainty bounds for a specific reliability parameter. The combined uncertainty parameters represent both the irreducible and event-driven issues that are associated with any given analysis case. This uncertainty contribution comes from the knowledge of the analyst integrating the information about material degradation and determining the likelihood of major pipe failure challenging the safety of a plan. As information about the pipe failure context and the knowledge about possible influences of RIM increases, the epistemic uncertainty decreases.

9.4.2. Material degradation and uncertainty analysis

Each methodology (DDM, PFM, I-PPoF) includes detailed considerations of uncertainty. For each application the uncertainty distributions are skillfully crafted, thus reflecting the knowledge and experience of the analyst and the overall material degradation state of knowledge. The latter aspect is always going to be challenged by peer reviewers and the material science experts; what are the justifications, how do assumptions impact the results, and are the results reasonable? The design and construction (fabrication, welding, installation) of piping systems respond to codes and standards requirements, regulations and specific operating environments. An uncertainty distribution reflects piping design information, material properties and empirical data as shown in Fig. 40.

A process for translating the abstract uncertainty concepts into practical guidelines for piping reliability analysis begins with the fundamentals. That is, deep knowledge of piping design principles,

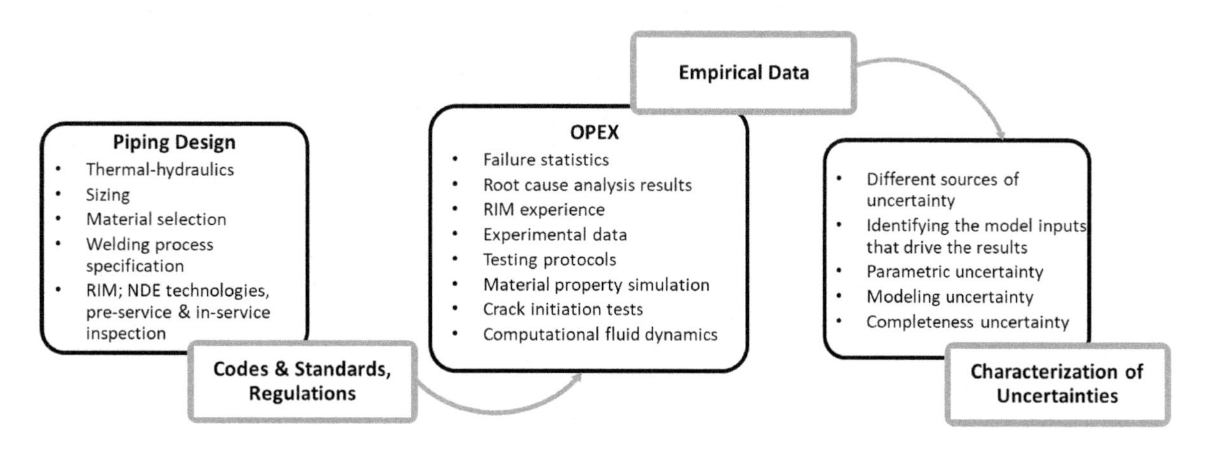

FIG. 40. Piping materials uncertainty characterization process.

material degradation, and the empirical data associated with piping in the different operating environments. The treatment of uncertainty differs according to the type of piping system for which probabilistic failure metrics are estimated. There are four types of piping systems:

(1) *Primary system piping connected to the reactor pressure vessel.* The predominant materials are stainless steels, carbon steel clad with stainless steel and nickel base alloys. The conjoint requirements for material degradation are well known. Nevertheless, pipe flaws sometimes develop in locations believed to be immune to degradation.
   (a) Parametric uncertainty relates to the uncertainty in the frequency of crack initiation and propagation. This uncertainty is influenced by welding processes, material selection, nondestructive evaluation (NDE) techniques (e.g. extent and type of surface preparation for NDE) and stress improvement processes. It is also influenced by the known degradation susceptibilities of different materials. For example, the shape of the uncertainty distribution is different for an evaluation boundary with ample field experience data as opposed to experimental data only [24].
   (b) Modelling uncertainty means the uncertainty in the conditional failure probability. Different models yield different answers, potentially with significant impact on the 5th, 50th and 95th percentiles of the pipe failure frequency. The modelling of long term effects of high temperature and high pressure on fracture toughness could have a significant impact on the shape of an uncertainty distribution. The modelling uncertainty could be an important influence on results interpretation when a probabilistic acceptance criterion applies.
   (c) Completeness uncertainty addresses whether an analysis accounts for all material degradation scenarios. For example, could multiple degradation mechanisms interact to cause a slow or rapid crack propagation? Access to relevant empirical data is important, and physics-of-failure models can provide valuable insights into the completeness uncertainty.
(2) *Piping connected to the primary system piping, providing chemistry control, level control, pressure control, safety injection, and residual heat removal, and ensuring containment integrity.* The predominant materials are stainless steels and nickel base alloys. The piping consists of large, medium and small diameter piping that are exposed to different operating environments (e.g. intermittent flow and stagnant flow conditions). Some sections of the piping are within the primary system pressure boundary; the majority of the piping is outside of the primary pressure boundary. Depending on the evaluation boundary, different RIM processes apply.
   (a) The parametric uncertainty is influenced by the functionality of a system as well as the routeing of piping for which probabilistic failure metrics are sought (i.e. local flow conditions).
   (b) The modelling uncertainty accounts for pressure, temperature and flow conditions as well as the type of surveillance testing which is performed to verify operability of standby safety systems. The configuration of an evaluation boundary could determine whether a susceptibility to thermal fatigue exists [24].
   (c) For this class of systems, the completeness uncertainty should acknowledge possibilities for gas intrusion and voiding in sections of piping normally isolated from the primary pressure boundary. Less than adequate post-maintenance and surveillance testing procedures could make piping susceptible to over-pressurization given an automatic system actuation signal. A water hammer load could exceed allowable stresses.
(3) *Support system piping that provides cooling water to safety related equipment, spent fuel pool cooling and fire protection water supply.* The plant-to-plant variability in the piping system designs is significant. The RIM process implementations vary significantly as well. The predominant materials are carbon steel and stainless steel, but high alloy stainless steel and high density polyethylene materials are also used.
   (a) The parametric uncertainty is addressed through a screening of the empirical data to account for different operating environments and plant-to-plant variability in piping system designs.

(b) Model uncertainty means that different conditional failure probability models are used to reflect variability in the process medium corrosion potential, low operating temperature and pressure, and multiple degradation mechanisms.
(c) Completeness uncertainty includes consideration of the conjoint requirements for degradation absent from any OPEX but with availability of limited experimental data. Susceptibilities to hydraulic transients such as water hammer could have a very significant impact on the uncertainty in the calculated probabilistic failure metrics.

(4) *Balance of plant piping systems that are needed for power conversion (i.e. high pressure and temperature piping for converting thermal energy to mechanical and electric energy).* Advanced WCRs use carbon steels selectively, and low alloy steels and stainless steels in wet steam environments. The empirical data for the different combinations of materials and operating environments are very extensive. RIM processes have been especially developed for these piping systems.
(a) As for the support system piping, the parametric uncertainty is addressed through a detailed screening of the empirical data to account for different operating environments and plant-to-plant variability in piping system designs.
(b) Model uncertainty is addressed using conditional failure probability models developed for the different combinations of materials and process media (dry steam, wet steam, condensate, feedwater).
(c) For the completeness uncertainty, extensive computer simulation data exist that correlate different piping geometries and operating environments with the susceptibility to flow assisted degradation. Water hammer loads acting on piping locations that are not susceptible to flow assisted degradation have a strong impact on the shape of uncertainty distributions. In advanced boiling water reactors, the use of hydrogen water chemistry could enhance the propensity for flow assisted degradation.

10. DOCUMENTATION

10.1. OBJECTIVES

The objective of step 7 of the analysis framework is to produce a traceable description of the processes used to develop the quantitative assessments of piping reliability. The objective of an analysis determines the depth of the documentation to be provided. Regulatory requirements and guidelines and relevant codes and standards determine or influence the scope and depth of the documentation. The users of the analysis results should be defined, including requirements for peer review. Applicable regulatory requirements should be identified. The analysis documentation should be organized to ensure that the information and data are scrutable, which means that the assumptions, input/output data, models selected, etc. are clearly stated.

10.2. DOCUMENTATION GUIDELINES

The processing of the information in step 6 (synthesize the insights and results) includes the output of step 6 and the cumulative output of all previous steps of the analysis framework. The documentation is to give a traceable record of the analysis; from step 1 through to step 6. Proposed documentation guidelines are found in Tables 16 and 17.

TABLE 16. SUGGESTED MINIMUM CONTENT OF DDM ANALYSIS DOCUMENTATION

Item	Item	Suggested minimum content to facilitate effective review
1	DDM software	In cases that a sufficiently similar software is not available to perform a meaningful benchmarking comparison, propose a way to give sufficient access to the software, for example, through an informal review meeting
2	Models	Document the model or models to a sufficient level of detail that a competent analyst already familiar with the relevant subject area could independently implement the model(s) Provide a basis for all significant aspects of the model(s), including why the selected models are sufficiently reliable for the intended application, with identification of important uncertainties or conservatisms Document any algorithms or numerical methods needed to implement the model(s) Discuss any significant assumptions, approximations, and simplifications, including their potential impacts on the analysis
3	Inputs	Document the inputs in detail, including specifying their values, including the statistical distributions used to characterize the uncertainties Provide the basis for the input values used, including why the input basis is considered sufficiently reliable for the application Document the process for retrieving and classifying the OPEX data Document the basis for pipe failure rate exposure terms Present the method and basis for treating epistemic and aleatory uncertainties
4	Input importance and sensitivity studies	Document an assessment of input importance, with the objective to identify the subset of inputs that have the greatest impact on the analysis results or conclusions Document the basis for how the prior distributions are defined. The documentation should include an assessment of the sensitivity of results to the choice of prior distribution
5	Verification and validation	Identify any applicable quality assurance programme, plan and/or procedures, as well as the quality assurance standards met Include a basic description of the measures for quality assurance, including verification and validation of the DDM software as applied in the subject report Document any benchmarking activities performed for the DDM software
6	Uncertainties	Summarize the overall Monte Carlo sampling structure (or other probabilistic treatment) and simulation framework, including their basis Include a summary discussion of key uncertainties or biases stemming from assumptions and simplifications to make real-world phenomena tractable, based, at a minimum, upon qualitative assessment
7	Acceptance criteria	Document the probabilistic acceptance criteria and their bases

TABLE 17. SUGGESTED MINIMUM CONTENT OF PFM ANALYSIS DOCUMENTATION [158]

Item	Item	Suggested minimum content
1	Information made available to reviewer	The analyst should have a plan for making the PFM software and supporting documents available to enable the review of PFM analysis

TABLE 17. SUGGESTED MINIMUM CONTENT OF PFM ANALYSIS DOCUMENTATION [158] (cont.)

Item	Item	Suggested minimum content
1.1	PFM software	In cases that a sufficiently similar PFM code is not available for a reviewer to perform a meaningful benchmarking comparison, propose a way to give sufficient access to the PFM software, for example, through an informal review meeting
1.2	Supporting documents	As appropriate, the supporting technical and quality assurance documents and procedures for the PFM software should be available for examination by a reviewer in an in-person review meeting
2	Models	Document the model or models to a sufficient level of detail that a competent analyst already familiar with the relevant subject area could independently implement the model(s) Provide a basis for all significant aspects of the model(s), including why the selected models are sufficiently reliable for the intended application, with identification of important uncertainties or conservatisms Document any algorithms or numerical methods needed to implement the model(s) Discuss any significant assumptions, approximations and simplifications, including their potential impacts on the analysis
3	Inputs	Document the inputs in detail, including specifying their values and whether they are treated as deterministic or probabilistic (and if probabilistic, document the distribution from which the inputs are sampled) Provide the basis for the input values used, including why the input basis is considered sufficiently reliable for the application Document use of interpolation, extrapolation and truncation schemes, as well as curve fitting of data Document the approach for treatment of correlation or statistical independence of inputs, along with the corresponding basis for the approach Ensure that selected or sampled inputs remain consistent and physically valid if inputs are dependent on each other (e.g., due to physical processes) Present the method and basis for treating epistemic and aleatory uncertainties
4	Convergence	Explicitly demonstrate convergence for all temporal and spatial discretizations, as well as statistical convergence of the Monte Carlo simulation
5	Input importance and sensitivity studies	Document an assessment of input importance, with the objective to identify the subset of inputs that have the greatest impact on the analysis results or conclusions Revisit the most important inputs and discuss how the values or distributions for the most important inputs were confirmed to be treated appropriately
6	Verification and validation	Identify the applicable quality assurance programme, plan and/or procedures, as well as the quality assurance standards met Include a basic description of the measures for quality assurance, including verification and validation of the PFM software as applied in the subject report Include confirmation that the verification and validation cover the ranges of input values considered in the submittal Document any benchmarking activities performed for the PFM software

TABLE 17. SUGGESTED MINIMUM CONTENT OF PFM ANALYSIS DOCUMENTATION [158] (cont.)

Item	Item	Suggested minimum content
7	Uncertainties	Summarize the overall Monte Carlo sampling structure (or other probabilistic treatment) and simulation framework, including their basis Include descriptions of the pseudo-random number generation, sampling methods, sampling frequencies and applied spatial or temporal discretization Discuss the basis for any conservative treatments of input values or models Include a summary discussion of key uncertainties or biases stemming from assumptions and simplifications.
8	Acceptance criteria	Document the probabilistic acceptance criteria and their basis, for example, a previous established precedent. Document sensitivity studies that address how the 95th percentile of the failure frequency varies with assumptions associated with input variables

10.3. END USER EXPECTATIONS ON DOCUMENTATION

Included in this section are two examples of end user requirements on the documentation of piping reliability analysis results that are intended for probabilistic fitness for service analyses and PSA applications. An end user defines the scope of an analysis as well the format for how results are documented. A first example of end user requirements is the US Nuclear Regulatory Commission Regulatory Guide 1.245, a second example is the proposed Regulatory Guide DG-1382, a third example is the FRANX software developed by the Electric Power Research Institute and a fourth example is the R-Book developed by the Nordic PSA Group.

10.3.1. US NRC Regulatory Guide 1.178 Revision 2

Regulatory Guide 1.178 describes an approach to what is acceptable to the NRC for developing risk-informed in-service inspection (RI-ISI) programmes [159]. Developing an RI-ISI programme involves a determination of the pipe failure potential of individual pipe segments or welds within a piping system. The pipe failure potential is assessed using PFM or other methods. Section 2.1.5 of the Regulatory Guide 1.178 [159] states the following:

> "When implementing probabilistic fracture mechanics computer programs that estimate structural reliability and are used in risk assessment of piping, or other analytic methods for estimating the failure potential of a piping segment, some of the important parameters that the analysis needs to assess include the identification of structural mechanics parameters, degradation mechanisms, design limit considerations, operating practices and environment, and the development of a data base or analytic methods for predicting the reliability of piping systems. Design and operational stress or strain limits are assessed. This information is available to the licensee in the design information for the plant. The loading and resulting stresses or strains on the piping are needed as input to the calculations that predict the failure probability of a piping segment. The use of validated computer programs, with appropriate input, is strongly recommended in a quantitative RI-ISI program because it may facilitate the regulatory evaluation of a submittal. The analytic method should be validated with applicable plant and industry piping performance data.

To understand the impact of specific assumptions or models used to characterize the potential for piping failure, appropriate sensitivity or uncertainty studies should be performed. These uncertainties include, but are not limited to, design versus fabrication differences, variations in material properties and strengths, effects of various degradation and aging mechanisms, variation in steady-state and transient loads, availability and accuracy of plant operating history, availability of inspection and maintenance program data, applicability and size of the data base to the specific degradation and piping, and the capabilities of analytic methods and models to predict realistic results. Evaluation of these uncertainties provides insights to the input parameters that affect the failure potential and, therefore, require careful consideration in the analysis."

10.3.2. Structured method for preparing PFM analysis documentation

The US Nuclear Regulatory Commission has developed a graded approach to how to document the results of PFM analyses that are performed in support of decision making. As defined by the IAEA, a graded approach means "a process or method in which the stringency of the control measures and conditions to be applied is commensurate, to the extent practicable, with the likelihood and possible consequences of, and the level of risk associated with, a loss of control" [160]. This graded approach is described in Regulatory Guide 1.245 [161], with supporting technical information documented in NUREG/CR-7278 [162]. The proposed regulatory guide represents a regulator's perspective on how to document PFM studies, and it is an expansion of a proposed approach documented in an industry white paper from 2019 [158]. The contents of the proposed regulatory guide are as follows:

— Chapter 1: Introduction and background.
— Chapter 2: Contents of PFM submittal. This chapter includes a description of a structured approach and contents of PFM submittal.
— Chapter 3: Analytical steps in a PFM submittal, includes descriptions of the steps and actions that may be taken as part of developing a PFM study, in support of developing the contents of a PFM submittal.
— Chapter 4: Methods used in PFM analysis, detailed descriptions of individual statistical analysis methods that may be used in a PFM study, useful in performing the steps and actions described in NUREG/CR-7278 [162].

10.3.3. EPRI-FRANX software for internal flooding PSA

Internal flooding PSA (IF-PSA) refers to probabilistic treatment of flooding inside an NPP as a result of pipe failure [163]. A typical internal flooding PSA models a large number of flood scenarios and they account for all piping systems that potentially can contribute to internal flooding. Hundreds of flood scenarios are accounted for, each with a unique flooding initiating event represented by a pipe failure frequency. Developed by the Electric Power Research Institute, the FRANX computer software enables the definition of flood scenarios and the quantification of flood risk [164]. The program manages the list of scenarios that need risk calculations and manages the mapping of plant components to the elements of the PSA model that are affected by the scenario. Next, pipe failure frequency parameters are assigned to each scenario.

The internal flooding pipe failure frequency parameters are imported to the FRANX software as Excel files or CSV files that for each evaluation boundary lists the cumulative pipe failure frequency as a function of pipe size, type of operating medium, operating pressure, EBS and the estimated through-wall flow rates. The uncertainty distributions, or range factors (i.e. √95th/5th) are included. In this case, the data come from a source such as [165], and the software can calculate the cumulative pipe failure frequencies for user defined EBS and through-wall flow rates at a given operating pressure.

10.3.4. R-Book by Nordic PSA Group

The Nordic PSA Group was established as common forum for the discussion of issues related to PSA, with focus on the R&D needs. The group has funded a series of projects directly linked to the OECD/NEA database projects. In 2005, the group launched the 'R-Book' project to provide PSA practitioners with access to piping reliability parameters. A planning effort was launched with the following aims.

- Review pipe failure databases to identify the technical features that were considered important to the R-Book development.
- Review methods for piping reliability parameter estimation.
- Development, distribution and evaluation of a questionnaire that addressed user requirements on the planned R-Book (content, including level of detail and updating philosophy). Input from the international PSA community was sought.
- Development, distribution and evaluation of a questionnaire that addressed the availability and access to piping exposure term data (piping system design information including weld counts and pipe length information organized by system, size, material, process medium, safety classification).

The final version of the R-Book was published in 2011 (Fig. 41). Only an electronic version of the R-Book has been issued. It consists of a password protected CD with the project files structured as follows (also see Fig. 42):

- Summary report (WORD file and probability density function); Nordic PSA Group Report 04-007:01 [166, 167]. The report summarizes the data processing routines: from extracting event population data from the CODAP database, via definition of relevant exposure term data (i.e. the number of ROYs that produced certain event population times, the number of welds susceptible to a certain degradation mechanism), to the definition of calculation cases, and to the execution of a certain calculation.
- Theory manual. This document contains an overview of the general calculation format, including a technical basis for how to define exposure terms specific to a certain calculation case; for example, pipe failure in BWR reactor recirculation piping susceptible to intergranular SCC, failure of dissimilar metal reactor recirculation piping welds, etc.
- For each of the 26 systems, a file folder with the following system specific computer files:
  - Text file (WORD and probability density function formats) summarizing the underlying degradation mechanism analysis (degradation mechanism assessment results), input data (event population data and corresponding exposure term data), results summary and a discussion on how to use the R-Data results;
  - Data processing and results summary; Microsoft Excel files;
  - Oracle Crystal Ball reports for each calculation case; depending on the system, up to 24 individual Excel files. These 'reports' enable a user to not only reproduce a certain application but also to perform different types of sensitivity analyses depending on an intended application.

The total file size is ca. 90 Mb (compressed archive format), which is made up from a total of 96 independent and related files. All piping reliability calculations are based on OPEX data for the period 1970–2007.

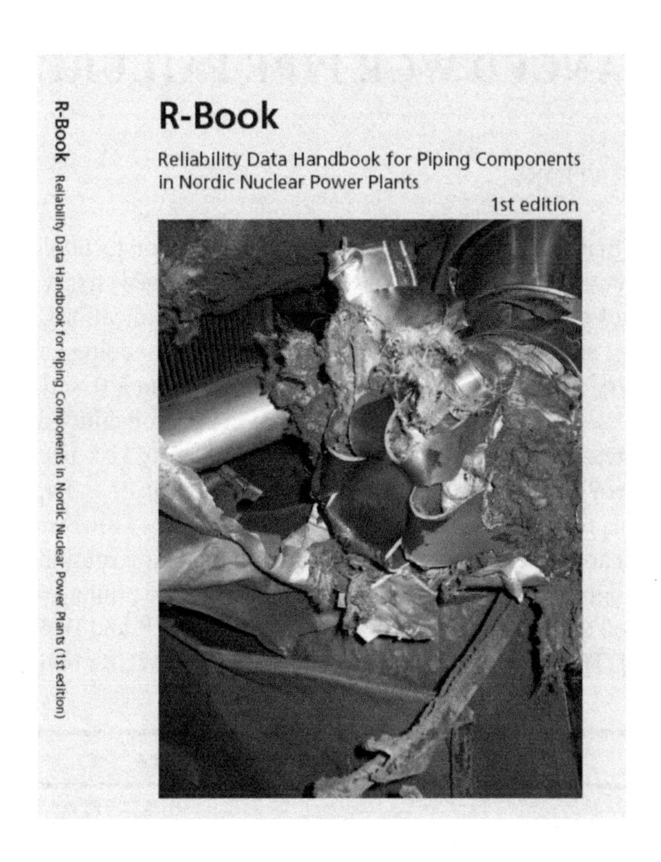

FIG. 41. R-Book report cover (courtesy of Nordic PSA Group).

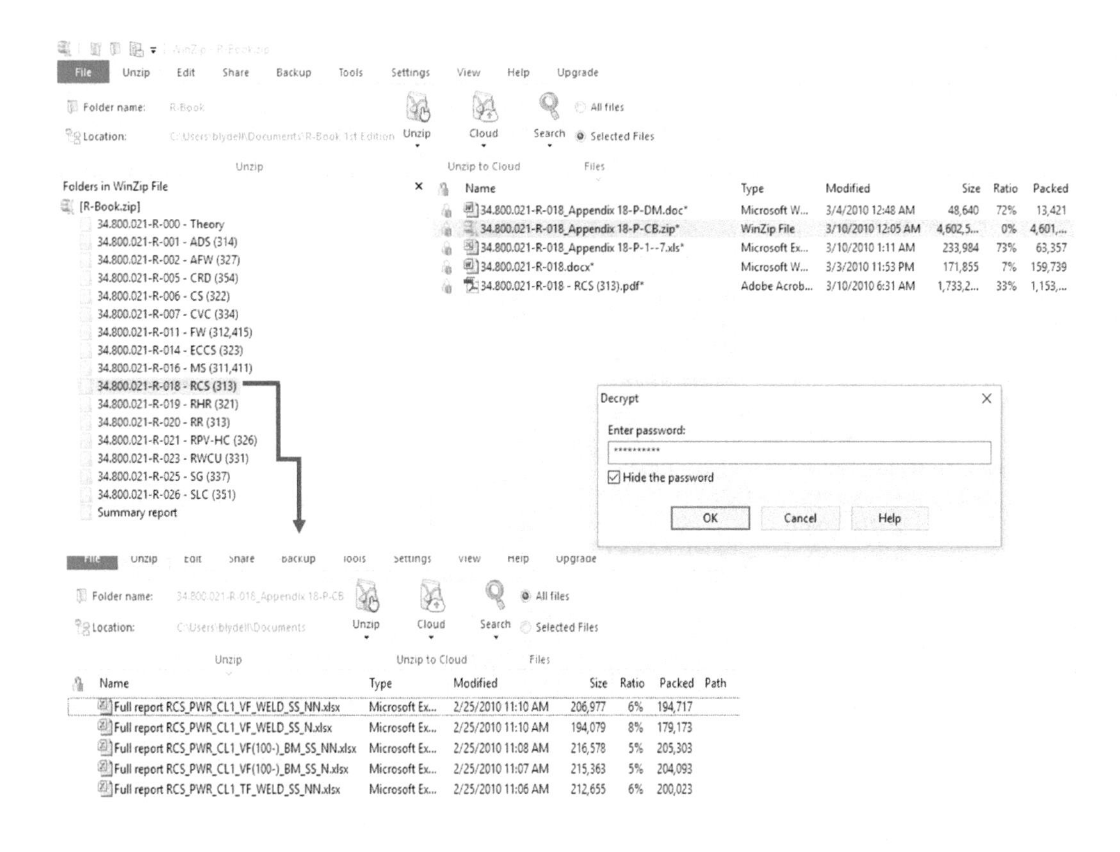

FIG. 42. R-Book content and structure.

11. ADVANCED WCR PIPE FAILURE RATES

11.1. BASIC CONSIDERATIONS

The methodologies that are described in this report are reactor technology neutral. They can be applied to any WCR and advanced WCR piping system reliability analysis. A lack of OPEX data could complicate the derivation of pipe failure rates for advanced WCR applications. The DDM approach may be viewed as inappropriate, and in applying the PFM or I-PPoF methodologies, confidence in the results may be questioned (i.e. very low failure frequencies much lower than 1.0×10^{-6} per ROY and location, and with large uncertainties). How much confidence is there in the probability density functions that have been developed to be representative of the piping structural integrity risk triplets: 'what can happen?', 'how likely is it to happen?', and 'what are the consequences if it does happen?'. The methods and techniques described in this report are applicable to advanced WCRs. However, the validation of results may require additional research or enhancements to existing calculation routines.

It is important to recognize what the differences are between the piping system designs of advanced WCR and WCRs. Details on the evolution of the BWR, CANDU, PWR and WWER designs can be found, for example, in [168–171]. The piping system designs for advanced WCRs build on lessons learned from

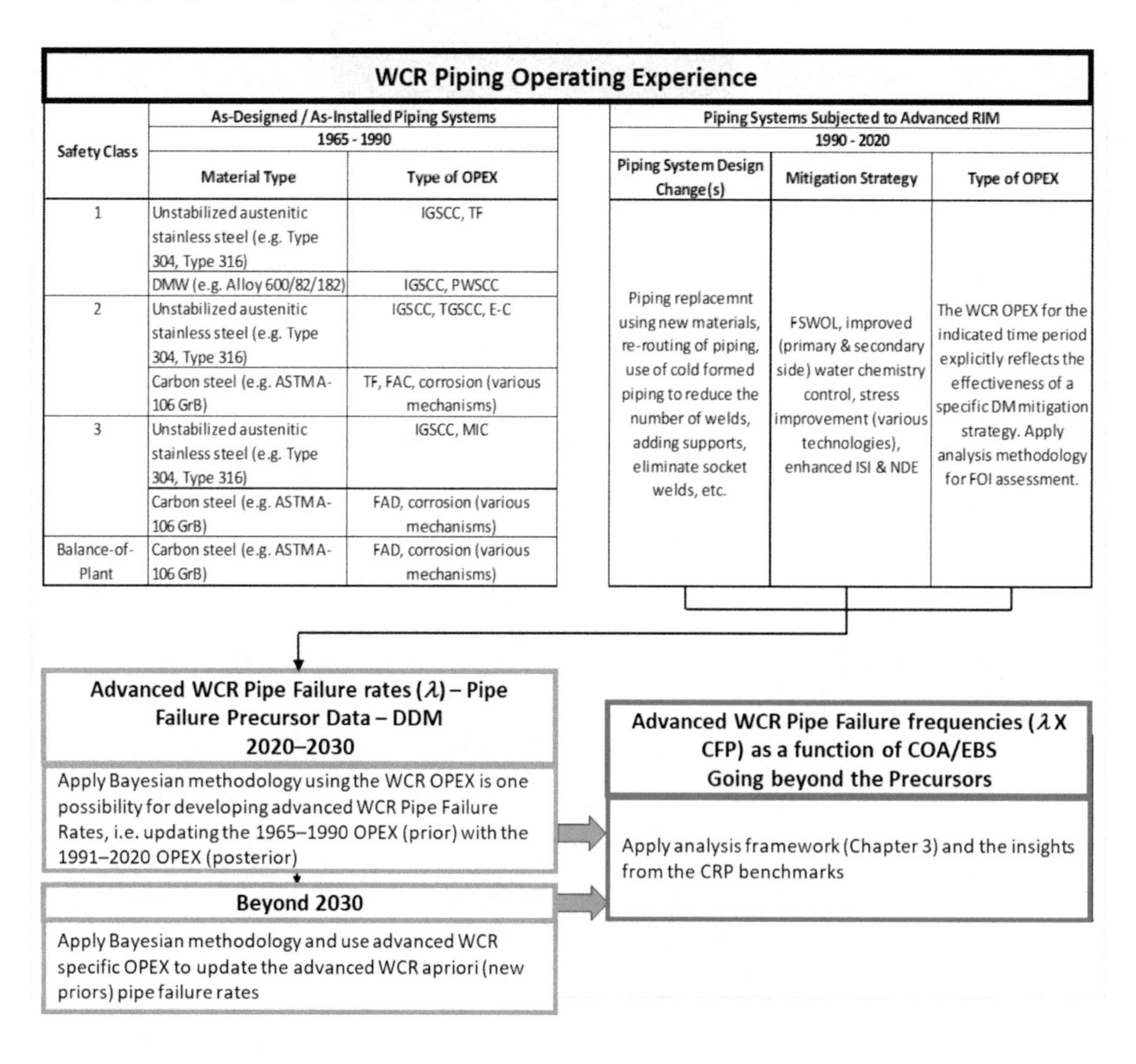

FIG. 43. Conceptual scheme for the derivation of advanced WCR centric pipe failure rates.

more than five decades of WCR design, commissioning and operation, and they acknowledge current codes and standards. Illustrated in Fig. 43 is a conceptual scheme for how to derive advanced WCR centric pipe failure rates (λ) and pipe failure frequencies, that conceptually correspond to $\lambda \times$ CFP[9]. The term 'centric' means that a failure rate reflects a specific set of advanced WCR operating environments, material properties and loading conditions. The conceptual scheme consists of elements 'A' through to 'F':

(a) *Material degradation insights*. The left-most part of the scheme in Fig. 43. The WCRs currently in operation were designed and commissioned in the 1960–1980 time frame. An extensive body of pipe failure data was generated during this period. The root cause evaluations, including non-destructive and destructive examinations, produced a deep understanding of the conjoint requirements for material degradation. Subsequent advances in material science and RIM technologies aided the development of strategies for how to mitigate piping material degradation in certain operating environments.
(b) *Proactive material degradation mitigation*. By the late 1980s plant operators had implemented major engineering changes to enhance the structural integrity of piping systems. These changes addressed the conjoint requirements for material degradation; improved water chemistry control, use of better materials and application of stress improvement techniques. For many WCR piping systems OPEX for the period 1990 to 2020 differs substantially from the period 1965 to 1990. These differences are well understood.
(c) *Learning process*. A Bayesian implementation of the DDM enables the derivation updated (posterior) pipe failure rates that acknowledge the material degradation 'learning effects'. For example, the extensive pre-1990 OPEX can be used to develop prior pipe failure rate uncertainty distributions that are updated using the post-1990 OPEX [172]. Conceptually simple to do, but nevertheless should be formalized by applying the piping reliability analysis framework. Illustrated in Fig. 44 are the results of a before and after assessment of OPEX to determine the factor of improvement from implementing degradation mitigation processes. This analysis was done in support of the CRP and addresses intergranular SCC factor of improvements in a BWR primary system operating environment.
(d) *Advanced WCR prior pipe failure rates and pipe failure frequencies*. It is technically feasible to develop informed advanced WCR pipe failure rates on the basis of the existing WCR OPEX. The material types and RIM strategies that are considered for advanced WCR application are well vetted and proven to be effective in material ageing management. Applying the piping reliability analysis framework, a combination of methodologies, and using the insights gained from [24], should yield results that can withstand a formal peer review process.
(e) *Advanced WCR posterior pipe failure rates*. As the advanced WCR OPEX becomes available it becomes technically feasible to update the advanced WCR a priori failure rates. Or, to develop new prior pipe failure rate uncertainty distributions.
(f) *Advanced WCR pipe failure frequencies*. Implementing elements 'A' through to 'E' yields the input parameter for the development of advanced WCR pipe failure frequencies. The CRP benchmark report gives additional details on how to address the analysis of advanced WCR pipe failure rate.

11.2. APPLICABILITY OF WCR PIPING OPEX TO ADVANCED WCRs

This section discusses what differentiates the advanced WCR piping system designs from the WCR piping system designs. The first commercial WCR designs were developed in the mid-1950s. The WCR piping system designs evolved significantly over the next several decades and in response to new regulatory requirements, national codes and standards, and operational insights. A WCR commissioned

[9] In this section the term 'λ' represents pipe failure 'precursors' and the term '$\lambda \times$ CFP' represents the frequency of a pipe failure of certain magnitude. CFP is the conditional pipe failure probability.

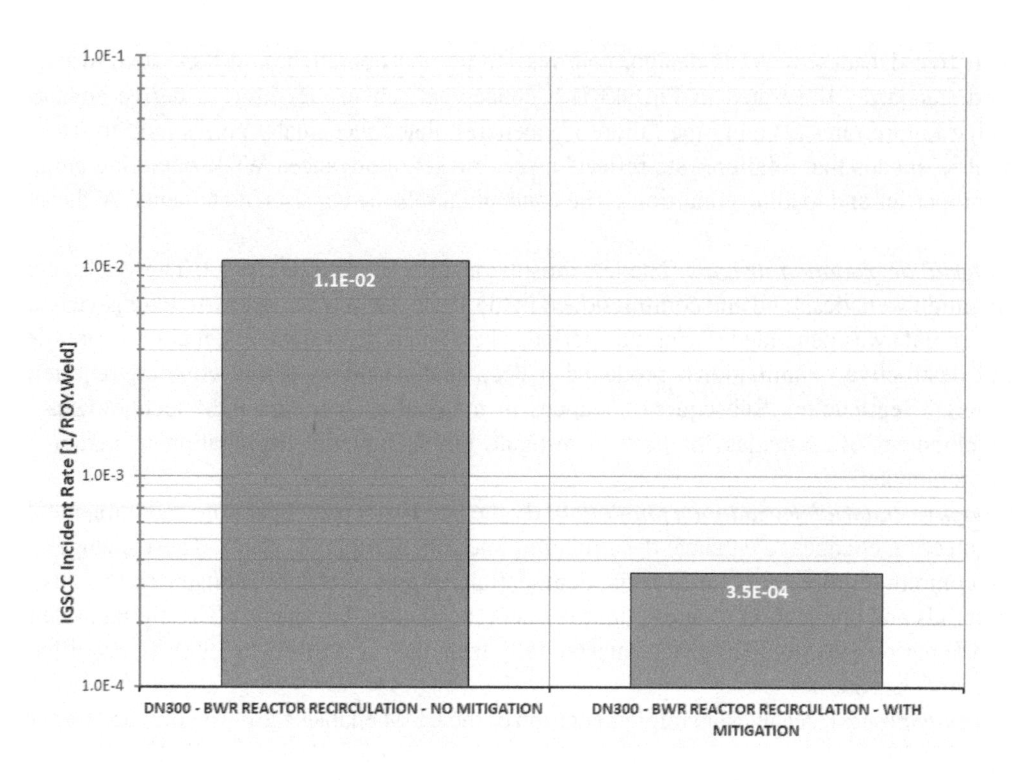

FIG. 44. Example of pipe failure rates before and after mitigation of degradation susceptibilities.

in the 1970s has undergone many piping system design changes, and the lessons learned from the WCR experience have been transferred to the advanced WCR piping system design philosophies:

— Less safety class 1 piping; fewer welds, improved access for inspection inside the containment:
  - Elimination of certain safety class 1 piping and large diameter reactor pressure vessel nozzles (e.g. no external reactor recirculation piping in Generation III/III+ BWRs). Elimination of piping between the steam generator and reactor coolant pumps in Gen III/III+ PWRs.
  - Fewer similar-metal welds in PWR RCS hot legs and cold legs.
  - Elimination of safety class 1 socket welds.

— Use of flow assisted degradation resistant balance of plant piping (e.g. high pressure steam piping).
— Improved support system piping designs.
— Use of corrosion resistant raw water piping material (e.g. high alloy stainless steel or high density polyethylene materials).

An overview of typical WCR and advanced WCR piping selections is given in Table 18. While some WCR plants operate with unmitigated and/or mitigated nickel base materials, advanced WCRs use SCC resistant nickel base alloys. Instead of carbon steels or 300 series austenitic stainless steels in raw water environments, it is projected that most advanced WCRs would have corrosion resistant stainless steel piping or plastic piping (e.g. high density polyethylene piping).

From the point of view of applicability of WCR piping OPEX to advanced WCRs, a notable observation concerns the effectiveness of degradation mitigation practices (see Tables 19 and 20) [93].

The existing WCR OPEX enables evaluations of piping degradation susceptibilities before and after a specific mitigation process or technology has been applied (i.e. a factor of improvement assessment).

TABLE 18. TYPICAL WCR AND ADVANCED WCR PIPING MATERIAL SELECTIONS

System group	Typical material selections	
	WCR	Advanced WCR
Primary coolant system	Stainless steel clad carbon steel (e.g. CE and B&W), cast austenitic stainless steels elbows, seamless austenitic stainless steel piping, Alloy 600/82/182 dissimilar metal welds	Cast austenitic stainless steels (ferrite content <20 FN), seamless austenitic stainless steel piping, Alloy 690/52/152 dissimilar metal welds
Emergency core cooling	300-series austenitic stainless steel	300-series austenitic stainless steel
Reactor auxiliary systems	Carbon steel, 300-series austenitic stainless steel	Carbon steel, 300-series austenitic stainless steel
Safety related raw water cooling system	Carbon steel, 300-series austenitic stainless steel, super-austenitic stainless steel	Super-austenitic stainless steel, HDPE (buried pipe sections)
Below ground service water and fire water system piping	Carbon steel, cast iron, HDPE	Stainless steel, super-austenitic stainless, HDPE
Feedwater and steam system piping	Carbon steel, low alloy steel, stainless steel)	Low alloy steel, stainless steel

11.3. APPLICABILITY OF THE ANALYSIS FRAMEWORK

The purpose of the piping reliability analysis framework is to ensure that the objectives of an analysis are translated into specifications for the required input parameters as well as for the quantification scheme. If done in sufficient detail, the unique advanced WCR piping reliability attributes and influence factors are defined such that OPEX data screening criteria are obtained. Possible but surmountable complications are envisaged when addressing new types of piping materials, including advanced structural alloys and non-metallic materials. High density polyethylene piping is increasingly being used in corrosive operating environments. The failure modes of plastic piping differ from metallic piping. This affects the methodology selection element of the CRP framework. PFM and I-PPoF would apply to plastic piping, assuming that due consideration is given to the unique high density polyethylene failure modes (e.g. creep and slow crack growth).

Figure 45 illustrates the results of an analysis to assess the expected performance of DN100 piping of different materials in a raw water environment. The plant system of concern is the service water system for which extensive WCR OPEX is available. A typical original material of choice was carbon steel, which is susceptible to microbiologically influenced corrosion in some raw water environments. In order to reduce the incident rate of corrosion failures, extensive piping replacements were made using 300-series stainless steel. However, this material proved to be more or less equally susceptible to microbiologically influenced corrosion as carbon steel. The analysis behind the results in Fig. 45 considered three analysis cases:

TABLE 19. EFFECT OF AGEING MANAGEMENT ON MATERIAL IN SCC AND FATIGUE SUSCEPTIBLE ENVIRONMENTS

Degradation mechanism		Material(s)	Operating environment(s)	WCR OPEX domain	Advanced WCR OPEX domain
					WCR OPEX domain/proactive ageing management
				Steps taken to manage/mitigate degradation (post-1980s)	Ageing management effect on OPEX
Stress corrosion cracking	External chloride induced SCC	Stainless steel — multiple grades	Multiple	Improved fabrication practices, cleanliness, pipe surface decontamination	Inconclusive (i.e. not possible to make clear distinction between the 'early-life OPEX' vs. long term OPEX)
	Transgranular SCC				
	Intergranular SCC	Unstabilized austenitic stainless steel Stabilized austenitic stainless steel Alloy 82/182	BWR primary water (hydrogen water chemistry, NMCA)	Improved water chemistry control, application of stress improvement process (e.g. MSIP, peening), use of SCC-resistant material, application of FSWOL, minimize number of welds)	Significant reduction noted in SCC incident rates
	Primary water SCC	Alloy 600/82/182	High temperature PWR primary water	Use of SCC-resistant material (e.g. Alloy 690/52/152), application of stress improvement process, application of FSWOL.	
fatigue	High cycle fatigue (small diameter lines)	Multiple	Multiple	Elimination of socket welds, fatigue monitoring, improved piping configuration, installation of vibration dampers	Inconclusive
	Low cycle fatigue	Multiple	Multiple	Improved piping configuration (e.g. added supports), enhanced welding technology	Inconclusive

TABLE 19. EFFECT OF AGEING MANAGEMENT ON MATERIAL IN SCC AND FATIGUE SUSCEPTIBLE ENVIRONMENTS (cont.)

Degradation mechanism	Material(s)	Operating environment(s)	WCR OPEX domain	Advanced WCR OPEX domain
				WCR OPEX domain/proactive ageing management
			Steps taken to manage/mitigate degradation (post-1980s)	Ageing management effect on OPEX
Thermal fatigue	Multiple	Hot/cold fluid mixing	Fatigue monitoring, improved non-destructive examination technology, improved system operating procedures, piping system re-configuration	'Somewhat' inconclusive
Corrosion fatigue	Multiple	High temperature environment	Re-configuration of piping (e.g. PWR Feedwater nozzle configuration)	Inconclusive (only limited OPEX exists, most of which are from the 1970s to early 1980s)
Environmental fatigue	Multiple	High temperature environment	Fatigue monitoring, improved plant operating procedures	Relationship between field experience and experimental and theoretical work yet to be determined

TABLE 20. EFFECT OF AGEING MANAGEMENT ON MATERIAL PERFORMANCE IN FLOW-ASSISTED DEGRADATION -SUSCEPTIBLE and CORROSIVE ENVIRONMENTS

Degradation mechanism		Material(s)	Operating environment(s)	WCR OPEX domain	Advanced WCR OPEX domain
					WCR OPEX domain/proactive ageing management
				Steps taken to manage/mitigate degradation	Ageing management effect on OPEX
Flow assisted degradation	Erosion-cavitation	Multiple	Multiple	Piping system reconfiguration, improved system operating procedures	Inconclusive — a function of the effectiveness of knowledge transfer
	Erosion-corrosion	Carbon steel	Multiple	Use of erosion-corrosion resistant materials	Inconclusive
	Flow assisted corrosion	Carbon steel, low alloy steel	High energy piping, single-phase or two-phase flow conditions	Use of flow accelerated corrosion resistant material, implementation of flow accelerated corrosion monitoring programme, improved secondary-side water chemistry	Significant reduction noted for flow accelerated corrosion induced degradation leading to leaks or ruptures
	Liquid droplet impingement erosion	Low alloy steel, stainless steel		Reconfiguration of piping, improved operating procedures/operational strategies (e.g. base load vs. load following)	Inconclusive; the relationships between extended power uprate and liquid droplet impingement erosion susceptibility to be determined.

TABLE 20. EFFECT OF AGEING MANAGEMENT ON MATERIAL PERFORMANCE IN FLOW-ASSISTED DEGRADATION -SUSCEPTIBLE and CORROSIVE ENVIRONMENTS (cont.)

Degradation mechanism		Material(s)	Operating environment(s)	WCR OPEX domain	Advanced WCR OPEX domain
					WCR OPEX domain/proactive ageing management
				Steps taken to manage/mitigate degradation	Ageing management effect on OPEX
Corrosion	Corrosion — general Crevice corrosion Dealloying — selective leaching Galvanic corrosion Microbial corrosion	Al-bronze, carbon steel, stainless steel (e.g. 300-Series)	Raw water (e.g. brackish, lake, pond, river, sea)	Chemical treatment of water, use of corrosion resistant material incl. non-metallic material, reconfiguration of piping, application of composite repair technology[a], installation of cathodic protection system, etc.	Inconclusive — however, high alloy austenitic stainless steels (e.g. AL-6XN, 254-SMO and 654-SMO) and high density polyethylene materials appear to perform very well in raw water environments

[a] For more information, see for example, https://adams.nrc.gov/wba/ (Accession No. ML20014E476; 30 to 96-inch Pipelines Upgrade Case Study: Navigating Safety Related Pipeline Upgrades with CFRP and ML20014E506, Carbon Fiber Reinforced Plastic (CFRP) Repair – Future Applications and R&D Gaps).

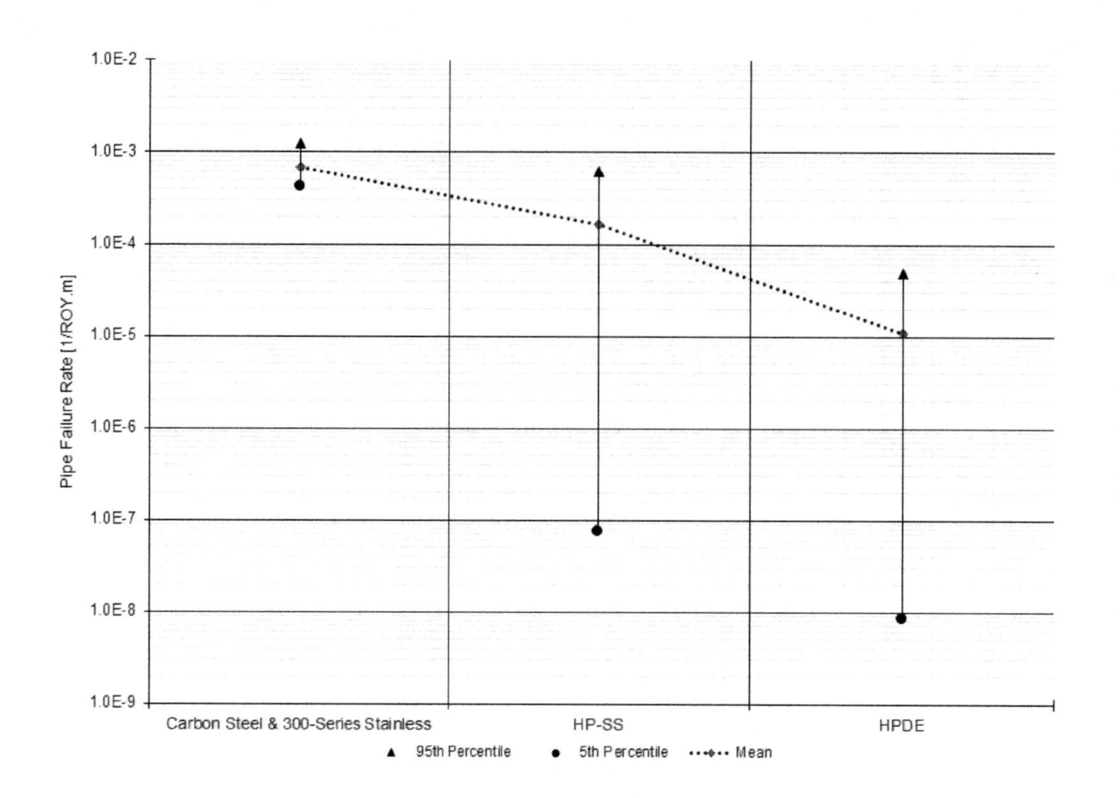

FIG. 45. Failure rates for DN100 piping in a raw water environment.

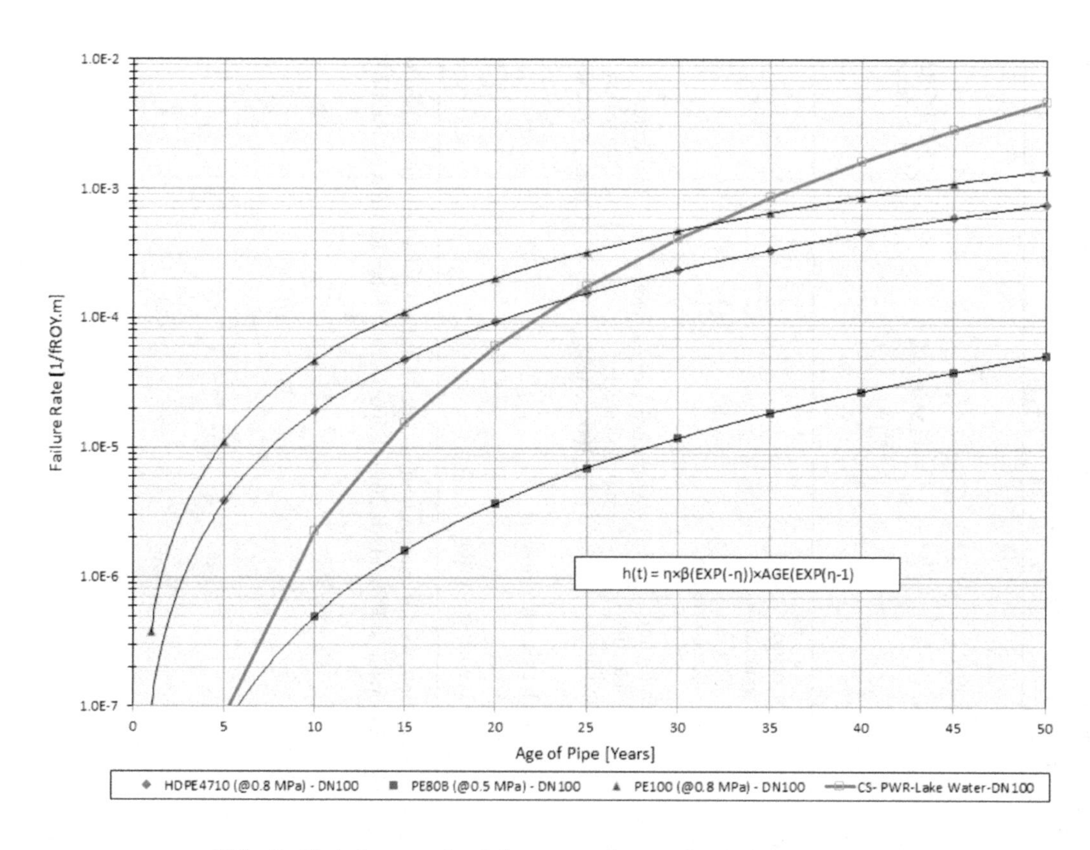

FIG. 46. High density polyethylene vs. carbon steel pipe failure rates.

(1) Carbon steel, DN100 piping in lake water environment. A DDM approach was used to derive pipe failure rates based on the US OPEX. This analysis is documented in [163].

(2) High alloy stainless steel materials were introduced in the late 1970s. Relative to the standard series 300 stainless, these materials have substantially higher chromium, molybdenum and nitrogen content. They are commonly referred to as high performance stainless steels because of their high resistance to crevice corrosion and pitting corrosion. An enhanced DDM approach was used to derive DN100 pipe failure rates for high performance stainless steel piping in a lake water environment. In this approach a mixture distribution probability matrix was developed to assign probability weights to different hypotheses about the corrosion resistance of high performance stainless steel. The methodology incorporated relevant OPEX data; Ref. [85] documents the methodology.

(3) At some WCR plants, continued issues with pinhole leaks, pitting, and other localized forms of pipe wall degradation due to microbiologically influenced corrosion have resulted in the replacement of portions of the original service water carbon steel piping with high density polyethylene piping. This material has demonstrated a high resistance to abrasion and biofouling and it is immune to general corrosion. Several advanced WCR plants use high density polyethylene material for safety and non-safety related piping including fire water system piping and service water piping. Reference [173] documents the results of an analysis which builds on laboratory test data to establish relationships between applied stress and failure time. A Weibull analysis of the test data yielded failure rates as a function of the age of the test specimens as shown in Fig. 46. These results were used to develop prior high density polyethylene pipe failure rates versus age and for different operating pressures.

11.4. STRATEGIES FOR UPDATING PIPE FAILURE RATES

In PSA, a question often raised is whether generic pipe failure rate estimates should be updated using plant specific OPEX data. The term 'generic pipe failure rates' refers to published, peer reviewed results that are available in the public domain. The large uncertainties and moderately low failure rates associated with piping can influence decisions so that NPP specific Bayes' updates are not usually performed. This is because usually there is insufficient NPP specific evidence to justify such a procedure. It has always been assumed that there would be only very small changes in pipe failure rate estimates if this type of Bayes' update were to be performed. In order to perform a technically sound Bayes' update of pipe failure rates the following questions arise:

— Is NPP specific data for failures and exposures collected and analysed in a manner consistent with the treatment of generic data in generic pipe failure rate estimates available within the public domain reports or industry reports?
— Is there a significant plant-to-plant or site-to-site variability in the pipe failure rate data that is reflected in the generic distributions? Figure 47 shows a wind rose diagram presenting the site-to-site variability in a raw water cooling system pipe failure experience.
— Whether NPP specific data need to be removed from the generic data in thus avoiding over counting the same evidence at two places. This is a generic issue in Bayes' updating with NPP specific data; it is usually ignored by assuming that the contribution to generic distributions from any specific NPP is small. This might not be true in the pipe failure rate case, especially if the NPP in question has an unusually high pipe failure incidence rate when compared to the rest in the industry.

To support this investigation, a limited analysis was performed to evaluate circulating water and service water pipe failure rates using US OPEX. The two systems were selected on the basis of their importance in some WCR internal flooding PSA studies. The hypothetical OPEX for PWR plants is compared with that of plant NPP 'X' for which plant specific estimates are sought. The results found in Tables 21 and 22 raise several questions regarding the process for data specialization. It is a topic for further research.

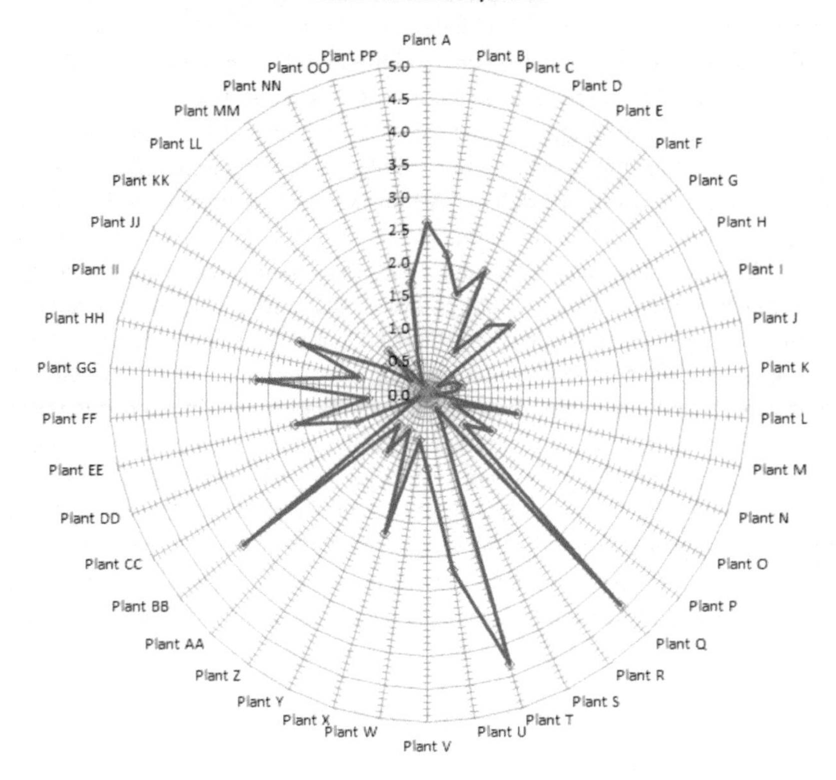

FIG. 47. Site-to-site variability in raw water cooling system pipe failure experience.

TABLE 21. SEPARATION OF PLANT 'X' FROM THE TOTAL PWR CIRCULATING WATER PIPE FAILURE EVENT DATA

Plant population	Failures	ROYs	Exposure term [ROY-m.]	Failure rate point estimate [1/ROY-m.]
All PWR plants	43	3960	376 586	1.1×10^{-4}
Plant 'X'	9	84	7988	1.1×10^{-3}
All less 'X'	34	3876	368 598	9.2×10^{-5}

TABLE 22. SEPARATION OF PLANT SPECIFIC DATA FROM TOTAL PWR SERVICE WATER PIPE FAILURE DATA ON SMALL LEAKS

Plant population	Failures	ROYs	Exposure term [ROY-m.]	Failure rate point estimate [1/ROY-m.]
Total PWR SW sea water	114	707	1 359 118	8.4×10^{-5}
Plant 'X'	20	84	161 479	1.2×10^{-4}
Total plant population less 'X'	94	623	1 197 639	7.9×10^{-5}

11.5. AGEING FACTOR ASSESSMENT PRINCIPLES

Opportunities to identify and evaluate temporal trends and potential effects of ageing on piping integrity are enabled by two important factors; (1) systematic and continuous OPEX data collection and evaluation, and (2) systematic and successive application of analysis methods and techniques to explore OPEX data in a formalized manner and to a common problem. It is not feasible to identify ageing trends on the basis of OPEX data alone. A statistical analysis may provide indications of possible trends in piping material performance, but it is a very challenging analytical task that requires access to a comprehensive database, which needs to be screened in order to address the many different factors that affect a data collection and analysis process; that is, accounting for changes in reporting routines and ageing management processes. Figure 48 depicts a snapshot of piping OPEX at the end of the calendar year 2020. It shows the number of pipe failures normalized against the plant population in a given plant age interval. Additional analyses are needed in order to obtain reasonable quantitative assessments of ageing factors, if any.

The next section includes a brief historical perspective on relevant research activities during the past 25 years. Also included are some results of a long term project to assess pipe failure rates and temporal trends. There are multiple technical approaches to the analysis of age dependent pipe failure rates. Before attempting a rigorous assessment, it is suggested that, as a first step, one should perform simple visual tests of graphical plots of failure data to obtain insights into temporal trends and possible ageing trends.

Identifying trends in pipe failure rates is a complex undertaking. Data homogeneity, data completeness and data analysis processes impact the insights regarding trends. The results as presented in Fig. 48 are affected as much by the data collection process, the completeness and quality of the data,

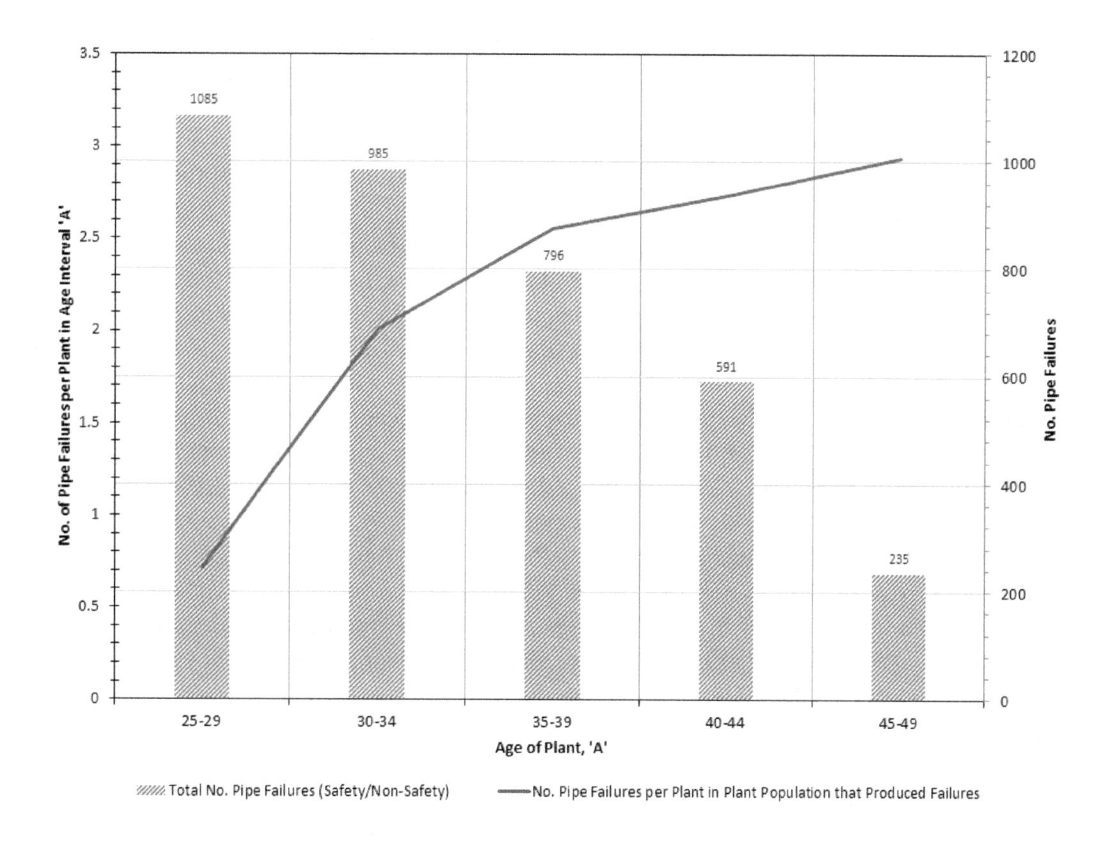

FIG. 48. Pipe failure trends[1].

[1] As of April 2021, the operating experience in this chart includes small, medium and large diameter piping of all safety classes as well as non-safety piping.

as they are affected by the technical approach that is used when binning the OPEX data (e.g. small versus large time intervals).

11.5.1. Research into ageing factor assessment

Research into materials ageing has been ongoing for about 50 years. Reviews of service experience data have been important to this research [174]. For example, in 1975 the US NRC established a first Pipe Crack Study Group charged with the task of evaluating the significance of SCC in BWR [86] and PWR plants [87]. Service experience reviews were an important aspect of the work by the Pipe Crack Study Group. As another example, major failures of condensate and feedwater piping (e.g. Trojan in 1985 and Surry in 1986) due to flow accelerated corrosion resulted in similar initiatives to learn from service experience and to develop mitigation strategies to prevent recurrence of flow accelerated corrosion induced pipe failures [88–91].

In the USA, the NRC sponsored the 'Nuclear Plant Aging Research' programme during 1985–1994 to collect information about ageing phenomena. Mainly, this programme collected a large body of qualitative information on plant ageing and the potential effects on plant safety. The 'Nuclear Plant Ageing Research' collected information has supported the formulation of the License Renewal Rule (10 CFR Part 54, 1995), and it has been utilized in subsequent NRC-sponsored ageing PSA feasibility studies.

In the report NUREG/CR-5378 [175], a methodology was presented for how to identify and quantify age dependent failure rates of passive components. Central to this approach was the detailed data analysis of component history data, including evaluation of failure trends. The chosen example is an auxiliary feedwater system and flow accelerated corrosion induced pipe failure. The calculated age dependent CDF remained virtually constant across the selected age span. It was a simplistic approach that since has been tried in other R&D efforts with similar results and insights. Adding basic events that represent passive component ageing effects tend not to produce any significant impact on the calculated risk metrics.

The report NUREG/CR-6157 [176] surveyed the NRC-sponsored work on the ageing of SSCs and the ageing PSA information sources. The report NUREG/CR-5632 [177] documented a proposed ageing PSA structure in which ageing effects are included in an existing PSA model through a so-called compound plug-in calculation. In this approach a selected ageing effect, flow accelerated corrosion in feedwater piping, was modelled by using a load-capacity probability calculation. Next, the flow accelerated corrosion piping failure event was inserted in a fault tree that modelled the loss of feedwater initiating event. The compound plug-in module computed the feedwater piping failure probability due to flow accelerated corrosion and as a function of operational time. The compound modelling concept is frequently used in PSA. It means that modelling is performed outside a PSA software platform and model integration is implemented during the PSA model quantification process.

In 2004, a Network on the Use of PSA for Evaluation of Ageing Effects to the Safety of Energy Facilities (EC JRC IE Ageing PSA Network) was established within the framework of the JRC FP-6 institutional Project No. 3131 Analysis and Management of Nuclear Accidents. Network meetings have been organized in Paris (2005), Bucharest (2006), Prague (2008) and Gösgen-Däniken (2010). The paper by Rodionov, Atwood, Kirchsteiger and Patrik [178] includes a summary of case studies that were performed within the framework of the Ageing PSA Network to demonstrate statistical approaches to quantify component ageing factors [179, 180]. The work of the Network concluded in 2010[10].

11.5.2. Temporal trends

The term 'temporal trend' describes the change of descriptive failure statistics (mean, upper/lower bound) over time. This may be due to ageing, such as a change in the physical properties of piping material (examples of thinning or cracking), or due to changing reporting routines and data collection processes. In

[10] For a full list of the APS Network reports, see: https://publications.jrc.ec.europa.eu/repository/search

other words, any temporal trends that are identified through a statistical analysis may primarily reflect the process that was implemented to collect OPEX data.

The term 'cohort effect', when used in piping reliability analysis, describes variations in observed structural integrity factors (such as onset of crack initiation and subsequent crack growth) as a function of operating time, plant age, plant design generation and degradation mitigation implementation strategies (e.g. full structural weld overlay, peening, induction heating stress improvement). The commercial NPP designs have evolved, and the reactor units designed in 1960s and commissioned in early 1970s exhibit quite different service experience histories than reactors designed and commissioned at later stages. Particularly strong cohort effects relate to intergranular SCC of BWR primary system unirradiated stainless steel piping. Those BWR units commissioned in the 1960s and 1970s had high intergranular SCC incidence rates; they have been attributed to an inadequate recognition of the relationships between the operating environments (water chemistry), material characteristics (carbon content) and stress conditions. The safety class 1 BWR intergranular SCC OPEX is summarized in Fig. 49. This OPEX was subdivided into five classes [181] (see Fig. 49):

(1) Intergranular SCC-1 (demonstration NPPs or Generation I) caused by inadequate water chemistry control during the early plant life are found in first generation BWRs. These reactor units were brought on-line in the 1960s.
(2) Intergranular SCC-2 (Generation II Early) caused by inadequate water chemistry control during the early plant life found in the second BWR design generation commercialized in early 1970s. Several units have entered into an extended period of operation.
(3) Intergranular SCC-3 (Generation II Midi) corresponds to the third BWR design generation. Multiple intergranular SCC-mitigation projects were implemented in the latter part of the 1980s.
(4) Intergranular SCC-4 corresponds to two specific BWR nuclear steam system supplier design generations (SWR69 and SWR72) that utilized stabilized austenitic stainless steel materials. This

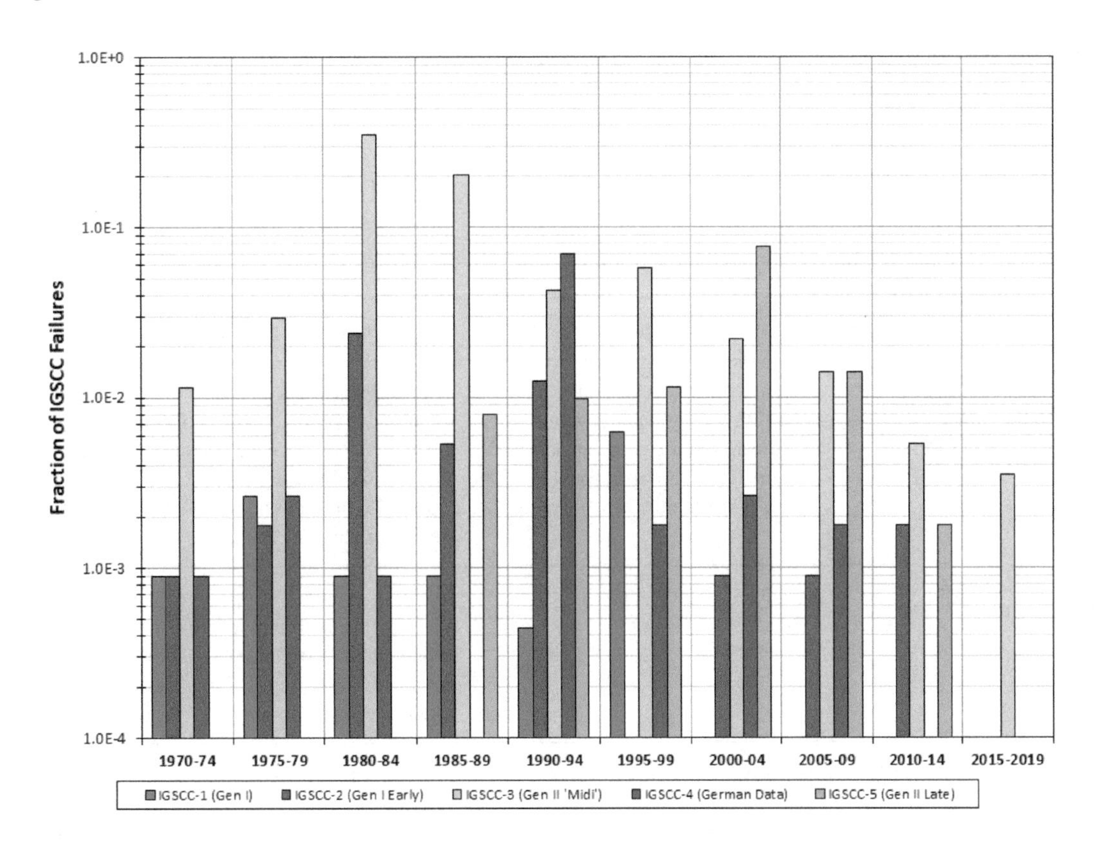

FIG. 49. Intergranular SCC incident rates.

class is limited to German NPPs for which intergranular SCC mitigation consisted of a re-design of the primary piping systems. This effectively eliminated the susceptibility to intergranular SCC.

(5) Intergranular SCC-5 (Generation II Late) corresponds to BWR designs for which intergranular SCC mitigation consideration was an integral part of the original piping design and NPP operating practices.

When performing successive analyses of OPEX data, an approach to the determination of temporal trends is to screen the data according to an assessed temporal change factor (TCF), which is defined as:

$$\mathrm{TCF} = \lambda_{(Pi)} / \lambda_{Ref} \tag{15}$$

where

$\lambda_{(Pi)}$ is the pipe failure rate for time period P_i;
λ_{Ref} is the pipe failure rate for a reference period (or base case).

Successive analysis means that a model of piping reliability is consistently applied to an evaluation boundary over a long time. Reference [182] presents the results of a multi-year (2005 to 2018) project to develop piping reliability parameters for use in internal flooding PSA. The project produced four editions of an internal flooding PSA piping reliability data handbook using a systematically applied DDM approach. The first edition used data collection on pipe failures observed during calendar years 1970 to 2004 (Period P1). Subsequent editions updated the previously derived pipe failure rates to account for new OPEX; Periods P2 (2008) through P4 (2015). These successive analyses enabled the evaluation of trends in the calculated pipe failure rates.

As a data screening tool, the temporal change factor concept can give insights into possible trends. The temporal change factor accounts for many different influences, including the potential change in material properties, changes in data collection processes, changes in reporting processes and requirements, changes in ISI requirements, etc., as shown in Table 23.

TABLE 23. SCREENING FOR ADVERSE TRENDS

TCF	Period(s)	Interpretation	Impact on pipe failure frequency
<1	P2, P3, P4 (1970–2020)	Effective flow accelerated corrosion ageing management. No significant trend change anticipated beyond 2020. FAC-free WCR piping performance is achievable	Applies to flow accelerated corrosion susceptible steam cycle piping systems. Extensive OPEX data available. Most flow accelerated corrosion susceptible WCR piping systems have been replaced with material resistant to flow assisted wall thinning
>1 but <2	All	No adverse trend noted in the OPEX	Insufficient data to support ageing factor assessments. Alternatively, existing ageing management programmes sufficiently effective to prevent adverse trends. Simple update; average across chosen time period
>2	P2 and P3 or All	Indicative of ageing of piping material	Results of formal ageing factor assessment could be factored into failure rate calculations

The development of methods for ageing factor analysis is motivated by the need to account for time dependent piping material degradation as NPPs enter into periods of extended or long term operation. The objectives of ageing factor assessments are to account for:

— *Ageing plant fleet.* The question to ask is: How effective is an existing ageing management programme with respect to long term material performance? Explicitly, a detailed identification and assessment of temporal trends would generate insights into the effectiveness of material ageing management and support to determine whether event recurrence patterns exist.
— *Renewal processes.* Piping systems are routinely replaced-in-kind or upgraded by replacing original material with material known or assumed to be resistant to degradation.
— *New OPEX data.* The OPEX with metallic passive components is continuously being updated. Are the data collection processes sufficiently complete to support quantitative ageing factor assessment?
— *Enhancements in RIM.* NPP life extension together with applications of non-destructive examination techniques and inspection qualification processes continue to evolve. Embedded in the OPEX data are the effects of non-destructive examination and changes in the reporting of pipe failures. Is the OPEX data that has been collected from plants in a period of extended operation (>40 years) a reflection of the improvements with respect to RIM implementation or could it be a result of unanticipated changes in material performance?

Results from the successive analyses offer insights regarding possible ageing factor assessment methodologies. The frequency of service water system pipe failures has exhibited a distinct trend over the lifetime of the operating fleet of PWRs and BWRs in the USA. For the first 20 years of plant life, the frequency of pipe failures in this system seemed to be fairly constant. The failure frequency appears to be significantly higher in the next 10 years of plant age and in the fourth decade the failure frequency is even higher.

The safety related service water piping system has the potential of producing flood sources for flood induced initiating events in an internal flooding PSA. This system is of interest in internal flooding PSA because it is found in many potential flood areas within a plant. The system has the potential to deliver large flood rates and flood volumes as a result of a pipe rupture.

11.5.3. Coarse pipe failure rate adjustment factors

Age dependent pipe failure rate estimation can be performed according to different analysis strategies to obtain piping reliability parameters as a function of the age of an affected pipe section at the time of its observed failure, or as a function of the temporal changes in the piping OPEX. An example of the latter analysis strategy is to calculate the pipe failure rate for different time periods that correspond to the different revisions of the internal flooding PSA piping reliability data handbooks developed by EPRI, Fig. 50 [182].

There are other factors that may contribute to the observed increases besides ageing. These include changes in inspection and reporting practices, implementation of the maintenance rule, and the fact that the OPEX database has been undergoing a continual update process. As reported in [182], many of the changes in inspection and reporting practice were confined to period 1 but changes that occurred over that 30 year period may have indeed tended to suppress the calculated average failure rates during those periods.

In order to make a coarse adjustment to the baseline internal flood frequencies to account for ageing effects, ageing factors can be obtained using a curve fit approach. A simplified technical approach to ageing factor assessment utilizes the calculated pipe failure rates for the periods P1−P4 (1970−2015). Each of the periods is representative of the accumulated OPEX against an average BWR and PWR fleet age.

The estimated pipe failure rates for each calculation case are used as anchor values when plotted against the average plant age for periods P1 through P3, and with additional calibration values calculated for selected time periods. In Microsoft Excel, a best-fit curve and equation are added to the four (P1−P4)

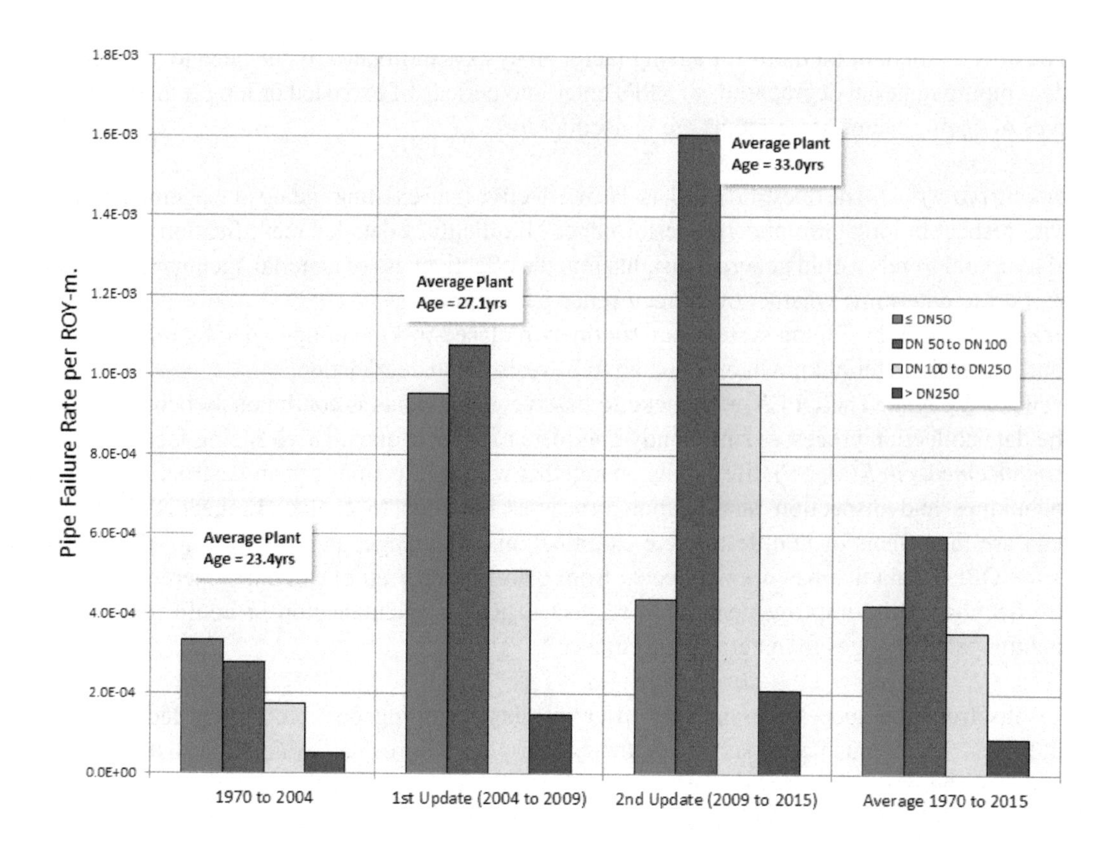

FIG. 50. Trends in pipe failure rates over three contiguous time periods.

calibration points, enabling the calculation of age dependent pipe failure rates. In this analysis a log-linear power model of ageing was used. The ageing factor was calculated as:

$$AF = \frac{\text{Age } "i"}{\text{Age } "1970\text{–}2020"} \tag{16}$$

where

$\text{Age } "i"$ is the pipe failure rate in time interval "*i*" (e.g. 2016–2020);
$\text{Age } "1970\text{–}2020"$ is the average pipe failure rate over the entire observation period.

Some plant specific PSAs have shown internal flooding to be a significant contributor to CDF [183]. Internal flooding is a special class of common cause initiating events. A pipe break can cause an initiating event (e.g. loss of system function) and impair the capability of multiple components and structures to perform safety functions due to submergence or water spray. Raw water cooling piping systems are of particular interest in internal flooding evaluations. The source of the cooling water is the ultimate heat sink (e.g. lake water, river water or sea water). Assessing the pipe break frequency as a function of break size or flood rate is a key step in the derivation of internal flood initiating event frequencies.

Assuming that an NPP has entered into long term operation, an analysis of age dependent pipe failure rates may be considered. Furthermore, assuming that internal flooding has been determined to be a major contributor to the CDF, the question is what type of data specialization should be considered for a pipe failure rate to be reflective of the operating environment, type of material and the age of the piping. Conditional on the availability of relevant pipe failure date, the following steps may be contemplated:

— Organize the data set by:
  - Pipe diameter.
  - Type of ultimate heat sink.
  - Type of raw water cooling system. The system may supply cooling water to a single heat load (e.g. component cooling water system) or multiple heat loads.
  - Event data.
  - Age of the pipe at the time of failure.

— Perform an initial trend analysis:
  - Bin the pipe failure population according to the year in which a failure was observed and in suitable intervals; for example, 3 year (1971–73, 1974–76, etc.) or 5 year intervals (1971–75, 1976–1980, etc.);
  - For each bin, compute the exposure term (number of ROYs × the length of the piping that is susceptible to material degradation);
  - For each bin, determine the average age of the plant population that produced the pipe failure population.

— For each bin and analysis case, calculate the temporal pipe failure rate;

— Use a Microsoft Excel line of best-fit through the scatter plots of failure rate vs. plant age and a log-linear power model to determine the scale and shape factors;

— For each calculation case determine the ageing factor AF.

There are additional analysis steps to consider. For example, some or perhaps several of the early life analysis bins may contain zero failures. This implies that a prior failure rate distribution is to be developed. An application of the AF approach to internal flooding initiating event frequencies also is to recognize the ageing management programmes that are in place to address piping material degradation.

12. CONCLUSIONS

12.1. STRATEGIES FOR ADVANCED WCR PIPE FAILURE RATES ESTIMATION

The seven-step piping reliability analysis framework is equally applicable to WCR and advanced WCR NPPs. The purpose of the framework is to ensure that the objectives of an analysis are translated into specifications for the required input parameters and the quantification scheme. The framework emphasizes the importance of using a consistent terminology and recognition of the end user requirements on an analysis with respect to the probabilistic failure metrics, documentation and quality assurance. When applied with sufficient rigour, the unique advanced WCR piping reliability attributes and influence factors are defined such that appropriate model input parameter selections and OPEX data screening criteria are obtained.

Possible but surmountable complications are envisaged when addressing new types of piping materials, including advanced structural alloys and non-metallic materials. High density polyethylene piping is increasingly being used in corrosive operating environments. The failure modes of plastic piping differ from metallic piping. This affects the methodology selection element of the framework.

A lack of OPEX data could complicate the derivation of pipe failure rates for advanced WCR applications. The DDM approach may be viewed as inappropriate, and in applying the PFM or I-PPoF methodologies confidence in the results may be questioned. In other words, very low failure frequencies, $\ll 1.0 \times 10^{-6}$ per ROY and location, and with large uncertainties. How much confidence is there in the probability density functions that have been developed to be representative of the piping structural integrity risk triplets becomes a question. The methods and techniques that are described in this publication are applicable to advanced WCRs. However, the validation of results may require additional research or the development of modifications to existing calculation routines.

It is important to recognize the differences that exist between the piping system designs of advanced WCRs and the preceding reactor designs. The piping system designs for advanced WCRs build on the lessons learned from more than five decades of WCR operation, and they reflect current codes and standards, and new RIM processes. A conceptual scheme is presented for how to derive advanced WCR centric pipe failure rates. The term 'centric' means that a failure rate reflects a specific set of advanced WCR operating environments, material properties and loading conditions. This scheme has multiple elements described as follows:

(a) *Material degradation insights*. The WCRs that are currently in operation were designed and commissioned in the 1960–1980 time frame. An extensive body of pipe failure data was 'generated' during this period. Subsequent advances in material science and RIM technologies aided the development of strategies for how to mitigate piping material degradation in certain operating environments.
(b) *Proactive material degradation mitigation*. By the late 1980s plant operators had implemented major engineering changes to enhance the structural integrity of piping systems. These changes addressed the conjoint requirements for material degradation; improved water chemistry control, use of better materials and application of stress improvement techniques. For many WCR piping systems, OPEX for the period 1990–2020 differs substantially from the period 1965–1990. The differences are well understood.
(c) *Learning process*. A Bayesian implementation of the DDM supports the derivation updated (posterior) pipe failure rates that acknowledge the material integrity learning effects. For example, the extensive pre-1990 OPEX can be used to develop prior pipe failure rate uncertainty distributions that are updated using the post-1990 OPEX. Conceptually simple to do, but nevertheless should be formalized by applying the piping reliability analysis framework.
(d) *Advanced WCR prior pipe failure rates*. It is technically feasible to develop informed advanced WCR pipe failure rates on the basis of the existing WCR OPEX. The material types and RIM strategies that are considered for ADVANCED WCR application are well vetted and proven to be effective in material ageing management.
(e) *Advanced WCR posterior pipe failure rates*. As the advanced WCR OPEX becomes available it becomes technically feasible to update the advanced WCR a priori failure rates. Or, to develop new prior pipe failure rate uncertainty distributions.
(f) *Advanced WCR pipe failure frequencies*. Implementing elements 'A' through to 'E' yields the input parameter for the development of advanced WCR pipe failure frequencies. The CRP benchmark report gives additional details on how to address the analysis of advanced WCR pipe failure rate.

12.2. INSIGHTS FROM THE APPLICATION OF DIFFERENT METHODS

The IAEA-TECDOC-1988 [24] documents the insights into the applicability of different methodologies that are a base guide when applying them for piping reliability analysis of the advanced WCRs:

— Implementations of the three families of methods can be computationally intense necessitating preplanning of the computations, and the post-processing of results tends to be time-consuming.
— Regardless of the selected methodology, the successful implementation depends on the experiences that had been obtained by the analysts from previously performed practical applications.
— The value and importance of OPEX data and experimental data could not be overstated as all methods benefit from and require some form of information on material performance and pipe failures.

— The PFM suffers from the tail sensitivity problem, which means that the computed probabilistic failure metrics are sensitive to very uncertain input parameters. The DDM cannot be applied to any problem especially defined to be solved on a fracture mechanics premise. In other words, the objectives of an analysis are conditioned on the context of an analysis (i.e. whether an analysis should be risk-informed and be input to a PSA task or is in the context of a fitness for service analysis). Therefore, the insights obtained from the inter-comparison of the three methods as documented in [24] represent a link to developing a modelling approach that uniquely responds to the advanced WCR analytical challenges.
— The lack of OPEX necessitates the interpretation and analysis of laboratory test data. A crucial point for the mechanistic approaches is the compilation of relevant references with respect to, for example, Alloys 690/52/152 crack initiation and crack growth rates. This clearly demonstrates the complexities of predicting the probabilistic failure metrics.

In advanced WCR piping reliability analysis there are two different potential analytical challenges facing an analyst: (1) an analysis boundary consisting of a material or combination of materials for which OPEX is obtainable including information on the applicable RIM strategies, and (2) an analysis boundary for which there is no known failure history. The first of these challenges can be addressed using any of the three methodologies and given that their implementation accounts for the effect of an RIM strategy that has been approved for advanced WCR use. The second challenge can also be addressed using any of the different methodologies. The effort needed to develop necessary input parameters can be significant, however. The integrated modelling of the many uncertain input parameters requires a carefully crafted implementation of the uncertainty analysis.

12.3. MAIN CONCLUSIONS

Many different piping reliability methods have been proposed, and some of these have been subjected to substantial enhancements that respond to the technical insights that have been obtained from a broad range of practical applications. However, there are methods that perform well when applied in a PSA setting. There are other methods specifically developed for fitness for service applications. Synergies across the different methods do exist as summarized in Fig. 51.

What matters the most is the way a certain method is selected and applied, the validity of the computational tools, the quality of the processes for developing input parameters, and the treatment of uncertainties. An implementation of the analysis framework promotes a consistency and transparency in the organization of an advanced WCR piping reliability analysis effort.

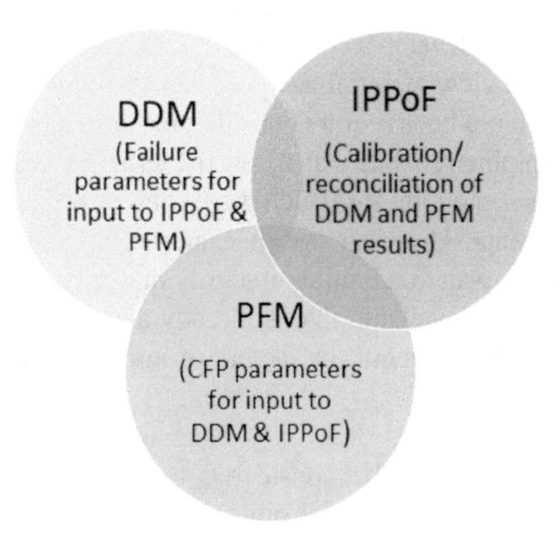

FIG. 51. Synergies among different methodologies.

The definition of what constitutes a pipe failure is a critical aspect of piping reliability analysis. There is ambiguity in the use of pipe failure mode terminology. Fundamentally, the term 'failure' implies that the integrity of a pressure boundary is compromised. The manner in which pipe material degrades and the consequence of a degraded condition influence how structural reliability is modelled.

As an example, in PSA the definition of the consequence of a pipe failure is very important. Hence, it is not meaningful to seek answers to questions like 'what is the frequency of a pipe leak?' or 'what is the frequency of pipe rupture?'. In probabilistic terms, the definition of failure versus its consequence affects the overall modelling approach, including the modelling of uncertainty. The consequence of a pipe failure can be characterized in terms of through-wall mass (kg/s) or volumetric flow rate (m^3/s) of process medium released into a confined or open space. The size of a through-wall pipe flaw can also be used and oftentimes is expressed in terms of an EBS, break opening area or crack opening area. The EBS is calculated using engineering analyses that involve consideration of fluid dynamics. In other words, it is the considerations of the consequence of a potential pipe failure that determines how it is modelled and how uncertainties are characterized.

12.4. CONTINUED RESEARCH

Continued research can fall into the following three categories: (1) expanding and refining the proposed piping reliability analysis framework through peer reviews that involve applications performed by different analysis, (2) methods development, and (3) piping reliability database development. Additional details are as follows:

(1) *Analysis framework*. Opportunities for expanding and refining the analysis framework can be identified through an international piping reliability analysis effort.
(2) *Methods development*. Each of the three categories of piping reliability methods are evolutionary. This means that each category has evolved in increments and iteratively, and in some cases active methods development has been under way for a long time. The mechanistic models can, in principle, be key to transfer operational experience to advanced WCRs. Similarly, the data-driven models have proven effective in terms of providing for a methodology to specialize WCR OPEX data to account for advances in material degradation mitigation as well as advances in ISI and non-destructive examination. Advanced WCR pipe failure rate assessment requires considerably more effort in terms of model validation and results interpretation than would be the case for WCRs for which there is extensive and well documented OPEX information. Further methods development is envisaged in order to make existing methods well qualified for future advanced WCR applications, and transferrable to a new generation of engineers and scientists. This can be achieved through an expert panel charged with the task to develop a coordinated research plan for the advancement of the piping reliability methodologies in view of anticipated regulatory and industry requirements.
(3) *Database development*. All methods require considerable amounts of input data and data processing efforts. In the context of piping reliability analysis, the term 'advanced WCRs' could be a misnomer in the sense that it implies something for which there is no or little data available; be it experimental data or field experience data. However, as an example, all operational WCRs have undergone significant upgrades with respect to reliability integrity management. A contention made is that there is sufficient WCR OPEX data available that is directly applicable to advanced WCRs in the sense that there exists deep knowledge about the factors of improvement that result from different RIM strategies.

A joint IAEA/NEA workshop can be considered to address how a database project like the OECD/NEA CODAP can support advanced WCR piping reliability analysis activities; more specifically: (1) how can an analyst benefit from CODAP in terms of developing the necessary inputs to an analysis, (2) what are the procedures and processes for ensuring that piping OPEX data becomes readily available to analysts and researchers, and (3) what procedural changes can be considered by the CODAP project to address current and future advanced WCR piping integrity issues.

REFERENCES

[1] GIBBONS, W.S., HACKNEY, S.D., Survey of Piping Failures for the Reactor Primary Coolant Pipe Rupture Study, GEAP 4574, Prepared for the U.S. AEC under Contract AT(04 3) 189, Atomic Power Equipment Department, General Electric Company, San Jose, CA (1964).

[2] UNITED STATES ATOMIC ENERGY COMMISSION, Int. Symp. on Fission Product Release and Transport under Accident Conditions (CONF 650407), Washington, DC (1965).

[3] BRAUN, C.F., & CO, A Review of Pipe Failure Experience, Technical Report 214, Prepared for the U.S. Atomic Energy Commission, Alhambra, CA (1969).

[4] BUSH, S.H., Reliability of piping in light water reactors, Nucl. Saf. **17** (1976) 568–579.

[5] BASIN, S.L., BURNS, E.T., Characteristics of Pipe System Failures in Light Water Reactors, NP 438, EPRI, Palo Alto, CA (1977).

[6] OECD NUCLEAR ENERGY AGENCY, Third Meeting of a Task Force on Problems of Rare Events in the Reliability Analysis of Nuclear Power Plants, NEA/CSNI(1978)51, Boulogne Billancourt, France (1978).

[7] KUTSU, O., Identification of Significant Problems Related to Light Water Reactor Piping Systems, ALO 89 (URS/Blume 7934), Sandia National Laboratories, Albuquerque, NM (1980).

[8] ALESII, G., HAYES, F.R., STANCAVAGE, P.P., Analysis of Scram Discharge Volume System Piping Integrity, NEDO 22209, Nuclear Energy Business Operations, General Electric, San Jose, CA (1982).

[9] HARRIS, D.O., LIM, E.Y., DEDHIA, D.D., WOO, H.H., CHOU, C.K., Fracture Mechanics Models Developed for Piping Reliability Assessment in Light Water Reactors, NUREG/CR 2301, U.S. Nuclear Regulatory Commission, Washington, DC (1982).

[10] BUSH, S.H., et al., Report of the U.S. Nuclear Regulatory Commission Piping Review Committee: Investigation and Evaluation of Stress Corrosion Cracking in Piping of Boiling Water Reactors, NUREG 1061, Vol. 1, Washington, DC (1984).

[11] KLECKER, R.W., BUSH, S.H., STROSNIDER, J., WICHMAN, K.R., Report of the U.S. Nuclear Regulatory Commission Piping Review Committee: Evaluation of Potential for Pipe Breaks, NUREG 1061, Vol. 3, Washington, DC (1984).

[12] BAGCHI, M., FAIR, J., HARTZMAN, M., O'BRIEN, J., SERKIZ, J., Report of the U.S. Nuclear Regulatory Commission Piping Review Committee: Evaluation of Other Loads and Load Combinations, NUREG 1061, Vol. 4, Washington, DC (1984).

[13] JANZEN, P, Piping Performance in Canadian CANDU Nuclear Generating Stations, AECL MISC 252, Chalk River Nuclear Laboratories, Chalk River, ONT, Canada (1984).

[14] HOU, S. N., BAGCHI, M., GUZY, D.J., MANOLY, K.A., O'BRIEN, J.A., Report of the U.S. Nuclear Regulatory Commission Piping Review Committee: Evaluation of Seismic Designs – A Review of Seismic Design Requirements for Nuclear Power Plant Piping, NUREG 1061, Vol. 2, Washington, DC (1985).

[15] SHAO, L.C., et al., Report of the U.S. Nuclear Regulatory Commission Piping Review Committee: Summary – Piping Review Committee Conclusions and Recommendations, NUREG 1061, Vol. 5, Washington, DC, 1985.

[16] BELICZEY, S., SCHULZ, H., The probability of leakage in piping systems of pressurized water reactors on the basis of fracture mechanics and operating experience, Nucl. Eng. Des. **102** (1987) 431–438.

[17] SCHULZ, H., Comments on the probability of leakage in piping systems as used in PRAs, Nucl. Eng. Des. **110** (1988) 229–232.

[18] HOLMAN, G.S., CHOU, C.K., Probability of Failure in BWR Reactor Coolant Piping, NUREG/CR 4792, U.S. Nuclear Regulatory Commission, Washington, DC (1989).

[19] BELICZEY, S., SCHULZ, H., Comments on probabilities of leaks and breaks of safety Related piping in PWR plants, Int. J. Press. Vessels and Piping **43** (1990) 219–227.

[20] HISER, A.I., MAYFIELD, M.E. (Eds), Proc. of the Seminar on Assessment of Fracture Prediction Technology: Pressure Vessels and Piping, NUREG/CP 0037, U.S. Nuclear Regulatory Commission, Washington, DC (1991).

[21] SWEDISH NUCLEAR POWER INSPECTORATE, Seminar on Piping Reliability, SKI Report 97:32, SKI, Stockholm, (1997).

[22] WILKOWSKI, G.M., OLSON, R.J., SCOTT, P.M., State of the Art Report on Piping Fracture Mechanics, NUREG/CR 6540, U.S. Nuclear Regulatory Commission, Washington, DC (1998).

[23] TREGONING, R., ABRAMSON, L. SCOTT, P., Estimating Loss of Coolant Accident (LOCA) Frequencies Through the Elicitation Process, NUREG 1829, U.S. Nuclear Regulatory Commission, Washington, DC (2008).

[24] INTERNATIONAL ATOMIC ENERGY AGENCY, Technical Insights from Benchmarking Different Methods for Predicting Pipe Failure Rates in Water Cooled Reactors, Final Report of a CRP, IAEA TECDOC 1988, IAEA, Vienna (2021).

[25] AMERICAN SOCIETY OF MECHANICAL ENGINEERS, Rules for Inservice Inspection of Nuclear Power Plant Components, Division 2, Requirements for Reliability and Integrity Management (RIM) Programs for Nuclear Power Plants, BPVC XI 2, New York (2019).

[26] OECD NUCLEAR ENERGY AGENCY, Workshop on PSA for New and Advanced Reactors, NEA/CSNI/R(2012)2, Boulogne Billancourt, France (2012).

[27] LYDELL, B., "Risk Informed Structural Integrity Management: Development of SMR Centric Piping Reliability Models", (Proc. ASME 2014 Small Modular Reactor Symposium), SMR2014 3328, ASME, New York, NY (2014).

[28] INTERNATIONAL ATOMIC ENERGY AGENCY, Corrosion and Erosion Aspects in Pressure Boundary Components of Light Water Reactors (Proc. Specialist Meeting Organized by the IAEA) IWG RRPC 88 1, IAEA, Vienna (1990).

[29] INTERNATIONAL ATOMIC ENERGY AGENCY, Material Degradation and Related Managerial Issues at Nuclear Power Plants (Proc. of a Technical Meeting), STI/PUB/1260, IAEA, Vienna (2006).

[30] INTERNATIONAL ATOMIC ENERGY AGENCY, Plant Life Management for Long Term Operation of Light Water Reactors: Principles and Guidelines, Technical Report Series No. 448, IAEA, Vienna (2007).

[31] INTERNATIONAL ATOMIC ENERGY AGENCY, Proc. of a Workshop on Erosion Corrosion Including Flow Accelerated Corrosion and Environmentally Assisted Cracking Issues in Nuclear Power Plants, Moscow, Russian Federation (2009).

[32] INTERNATIONAL ATOMIC ENERGY AGENCY, Approaches to Ageing Management for Nuclear Power Plants: International Generic Ageing Lessons Learned (IGALL) Final Report, IAEA TECDOC 1736, IAEA, Vienna (2014).

[33] INTERNATIONAL ATOMIC ENERGY AGENCY, Ageing Management and Development of a Programme for Long Term Operation of Nuclear Power Plants, Safety Standards Series No. SSG 48, IAEA, Vienna (2018).

[34] BURKE, W., LANCEFORD, W, SCOTT, P., DAVIDSAVER, S., FYFITCH, S., EPRI Materials Degradation Matrix, Revision 4, 3002013781, EPRI, Palo Alto, CA (2018).

[35] BUSH, S.H., "Statistics of pressure vessel and piping failures", Pressure Vessel and Piping Technology, A Decade of Progress, ASME, New York (1985) Ch. 8.9.

[36] MOREL, A.R., REYNES, L.J., Short term degradation mechanisms of piping, Nucl. Eng. Des. **133** (1992) 37–40.

[37] HURST, N.W. et al, A classification scheme for pipework failures to include human and sociotechnical errors and their contribution to pipework failure frequencies, J. Hazard. Mater. **26** (1991) 159–186.

[38] NEDZINSKAS, L., KLIMASAUSKAS, A., "Intergranular Stress Corrosion Cracking of Ignalina NPP Austenitic Piping of Outside Diameter 325mm", (Proc. Int. Conf. Strength, Durability and Stability of Materials, Klaipeda, Lithuania, 2003).

[39] BAEYENS, R. et al, Vibration and Noise Analysis in Nuclear Power Plants, EUR 5294e, Commission of the European Communities, Luxembourg (1974).

[40] MORITA, R., et al., Evaluation of acoustic and flow induced vibration of the BWR main steam lines and dryer, J. Nucl. Sci. Technol. **48** (2012) 759–776.

[41] MOORE, S., "A Review of Noise and Vibration in Fluid Filled Pipe Systems", (Proc. Acoustics 2016) Australian Acoustical Society, Toowong DC, QL, Australia (2016).

[42] OECD NUCLEAR ENERGY AGENCY, Operating Experience Insights into Electro Hydraulic & Instrument Air System Pipe Failures, CODAP Project Topical Report, NEA/CSNI/R(2015)6, Boulogne Billancourt, France (2015).

[43] OECD NUCLEAR ENERGY AGENCY, Operating Experience Insights into Pressure Boundary Component Reliability and Integrity Management, CODAP Project Topical Report, NEA/CSNI/R(2017)3, Boulogne Billancourt, France (2017).

[44] BRENNEN, C.E., Cavitation and Bubble Dynamics, Cambridge University Press, Cambridge, United Kingdom (2013).

[45] AZIZIAN, R., TORRADO, P., "A Practical Approach to Mitigate the Excessive Vibration of a Piping System

Subjected to Flow Induced Excitation", (Proc. ASME 2016 Pressure Vessels & Piping Conference), PVP2016 63088, ASME, New York (2016).

[46] MERZARI, E., et al., High Fidelity Simulation of Flow Induced Vibrations in Helical Steam Generators for Small Modular Reactors, LLNL JRNL 754259, Lawrence Livermore National Laboratory, Livermore, CA (2018).

[47] U.S. NUCLEAR REGULATORY COMMISSION, Pipe Wear due to Interaction of Flow Induced Vibration and Reflective Metal Insulation, Information Notice 2007 21, Supplement 1, Washington, DC (2020).

[48] GORDON, B.M., GORDON, G.M., "Corrosion in boiling water reactors", ASM Handbook, Vol. 13 (1987) 929–937.

[49] SEDRIKS, A.J., SCHULTZ, J.W., CORDOVI, M.A., Inconel Alloy 690 – A new corrosion resistant material, Corros. Eng. **28** (1979) 82–95.

[50] KHALIFEH, A., "Stress corrosion cracking damages", Failure Analysis (HUANG, Z–M., HEMEDA, S., Eds), IntechOpen, London (2019).

[51] KUMAR, S., RAMESH, T., ASOKKUMAR, K., Welding Studies on WB36 for Feed Water Piping, Int. J. Comput. Eng. Res. **6** 5 (2016) 36–52.

[52] AALTONEN, P. et al, Facts and Views on the Role of Anionic Impurities, Crack Tip Chemistry and Oxide Films in Environmentally Assisted Cracking, VTT Research Notes 2148, Technical Research Centre of Finland, VTT, Finland (2002).

[53] FORD, F.P., "Stress Corrosion Cracking of Carbon and Low Alloy Steels", Expert Panel Report on Proactive Materials Degradation Assessment, NUREG/CR 6923, U.S. Nuclear Regulatory Commission, Washington, DC (2007).

[54] SERKIZ. A.W., Evaluation of Water Hammer Occurrence in Nuclear Power Plants, NUREG 0927, Rev. 1, U.S. Nuclear regulatory Commission, Washington, DC (1984).

[55] VAN DUYNE, D.A., YOW, W., SABIN, J.W., Water Hammer Prevention, Mitigation, and Accommodation, Vol. 1: Plant Water Hammer Experience, NP 6766, EPRI, Palo Alto, CA (1992).

[56] ARASTU, A.H., LAFRAMBOISE, W.L., NOBLE, L.D. AND RHOADS, J.E., "Diagnostic Evaluation of a Severe Water Hammer Event in the Fire Protection System of a Nuclear Power Plant", (Proc. 3rd ASME/JSME Joint Fluids Engineering Conf.), ASME, FEDSM99 6891, New York (1999).

[57] BERGANT, A., SIMPSON, A.R., TUSSELING, A.S., Water Hammer with Column Separation: A Review of Research in the Twentieth Century, Department of Mathematics and Computer Science, Eindhoven University of Technology, Eindhoven, The Netherlands (2004).

[58] VAN DUYNE, D.A., MERILO, M. (Eds), Water Hammer Handbook for Nuclear Power Plant Engineers and Operators, TR 106438, EPRI, Palo Alto, CA (1996).

[59] LEISHEAR, R.A., Fluid Mechanics, Water Hammer, Dynamic Stresses, and Piping Design, ASME Press, ASME, New York (2013).

[60] CEUCA, S. C., Computational Simulations of Direct Contact Condensation as the Driving Force for Water Hammer, PhD Dissertation, Fakultät für Maschinenwesen, Technical University of Munich, Germany (2015).

[61] LEISHEAR, R.A., "Hydrogen Ignition Mechanisms for Explosions in Nuclear Facility Pipe Systems", (Proc. ASME 2010 Pressure Vessels & Piping Division Conf.), PVP2010 25261, ASME, New York (2010).

[62] U.S. NUCLEAR REGULATORY COMMISSION, "Thermal Aging Embrittlement of Cast Austenitic Stainless Steel Components", Staff Evaluation of License Renewal Issue No. 98 0030, Washington, DC (2000).

[63] HARRIS, D., QIAN, H., DEDHIA, D., COFIE, N., GRIESBACH, T., Nondestructive Evaluation: Probabilistic Reliability Model for Thermally Aged Cast Austenitic Stainless Steel Piping, 1024966, EPRI, Palo Alto, CA (2012).

[64] CHOPRA, O.K., Effects of Thermal Aging and Neutron Irradiation on Crack Growth Rate and Fracture Toughness of Cast Stainless Steels and Austenitic Stainless Steel Welds, NUREG/CR 7185, U.S. Nuclear Regulatory Commission, Washington, DC (2015).

[65] WÜTHRICH, C., Crack opening areas in pressure vessels and pipes, Eng. Fract. Mech. **5** (1983) 1049–1057.

[66] OLSON, R., Crack Opening Displacement, xLPR Models Subgroup Report, xLPR MSGR COD Version 1.0 (2016).

[67] KAPLAN, S., GARRICK, B.J., On the quantitative definition of risk, Risk Anal. **1** (1981) 11–27.

[68] BOURGA, R., MOORE, P., JANIN, Y. J., WANG, B., SHARPLES, J., Leak before break: Global perspectives and procedures, Int. J. Press. Vessels .Piping **129–130** (2015) 43–49.

[69] HECKMANN, K. ELMAS, M., SIEVERS, J., "Investigations on Various Leak Types from Operational Experience and Consequences for Leak Before Break Assessment of the Pressure Boundary", (Proc. 44th MPA Seminar, Stuttgart, Germany, 2018).

[70] OECD NUCLEAR ENERGY AGENCY, Committee on the Safety of Nuclear Installations: Recurring Events Volume 2, JT00149911, OECD, Paris (2003).

[71] FLEMING, K., LYDELL, B., Insights into location dependent loss of coolant accident (LOCA) frequency assessment for GSI 191 risk informed applications, Nucl. Eng. Des. **305** (2016) 433–450.
[72] OECD NUCLEAR ENERGY AGENCY, Updated Knowledge Base for Long Term Core Cooling Reliability, NEA/CSNI/R(2013)12, Boulogne Billancourt, France (2013).
[73] OECD NUCLEAR ENERGY AGENCY, Best Practice Guidelines for the Use of CFD in Nuclear Reactor Safety Applications, NEA/CSNI/R(2014)11, Boulogne Billancourt, France (2014).
[74] AMERICAN SOCIETY OF MECHANICAL ENGINEERS, Risk Informed Inspection Requirements for Piping, Nonmandatory Appendix R, ASME Boiler & Pressure Vessel Code, Section XI, Rules for Inservice Inspection of Nuclear Power Plant Components, New York (2019).
[75] INTERNATIONAL ATOMIC ENERGY AGENCY, Materials for Advanced Water Cooled Reactors, IAEA TECDOC 665, IAEA, Vienna (1992).
[76] ANDRESEN, P.L. et al, Expert Panel Report on Proactive Materials Degradation Assessment, NUREG/CR 6923, U.S. Nuclear Regulatory Commission, Washington, DC (2007).
[77] BOND, L.J., DOCTOR, S.R., TAYLOR, TT, Proactive Management of Materials Degradation – A Review of Principles and Programmes, PNNL 17779, Pacific Northwest National Laboratory, Richland, WA (2008).
[78] BUSBY, J.T., Light Water Reactor Sustainability: Materials Aging and Degradation Pathway Technical Programme Plan, ORNL/LTR 2012/327, Oak Ridge Natl Lab., TN (2012).
[79] BUSBY, J.T., Expanded Materials Degradation Assessment (EMDA), NUREG/CR 7153 Vol. 1, U.S. Nuclear Regulatory Commission, Washington, DC (2014).
[80] ANDRESEN, P. et al, Expanded Materials Degradation Assessment (EMDA): Ageing of Core Internals and Piping Systems, NUREG/CR 7153 Vol. 2, U.S. Nuclear Regulatory Commission, Washington, DC (2014).
[81] ODETTE, G.R., ZINKLE, S.J. (Eds), Structural Alloys for Nuclear Energy Applications, Elsevier, Cambridge, MA (2019).
[82] INTERNATIONAL ATOMIC ENERGY AGENCY, Ageing Management and Development of a Programme for Long Term Operation of Nuclear Power Plants, Specific Safety Guide No. SSG 48, IAEA, Vienna (2018).
[83] OECD NUCLEAR ENERGY AGENCY, Flow Accelerated Corrosion of Carbon Steel and Low Alloy Steel Piping in Commercial Nuclear Power Plants, CODAP Project Topical Report, NEA/CSNI/R(2014)6, Boulogne Billancourt, France (2014).
[84] JENKS, A., WHITE, G., BURKARDT, M., YOUNG, G., Materials Reliability Programme: Recommended Factors of Improvement for Evaluating Primary Water Stress Corrosion Cracking Growth Rates of Thick Wall Alloy 690 Materials and Alloy 52, 152, and Variants Welds, Technical Report 3002010756, EPRI, Palo Alto, CA (2017).
[85] LYDELL, B., FLEMING, K., Piping System Failure Rates for Corrosion Resistant Service Water Piping, Report No. 3002002787, EPRI, Palo Alto, CA (2014).
[86] U.S. NUCLEAR REGULATORY COMMISSION, Investigation and Evaluation of Cracking in Austenitic Stainless Steel Piping of Boiling Water Reactor Plants, NUREG 75/067, Washington, DC (1975).
[87] U.S. NUCLEAR REGULATORY COMMISSION, Investigation and Evaluation of Cracking Incidents in Piping in Pressurized Water Reactors, NUREG 0691, Washington, DC (1980).
[88] BROWN, E.J., Erosion in Nuclear Power Plants, AEOD/E4 16, Office for Analysis and Evaluation of Operational Data, U.S. Nuclear Regulatory Commission, Washington, DC (1984).
[89] CRAGNOLINO, G., CZAJKOWSKI, C., SHACK, W.J., Review of Erosion Corrosion in Single Phase Flows, NUREG/CR 5156, U.S. Nuclear Regulatory Commission, Washington, DC (1988).
[90] INTERNATIONAL ATOMIC ENERGY AGENCY, Corrosion and Erosion Aspects in Pressure Boundary Components of Light Water Reactors, Proc. Specialist Meeting Organized by the IAEA, IWG RRPC 88 1, IAEA, Vienna (1990).
[91] OECD NUCLEAR ENERGY AGENCY, Specialist Meeting on Erosion and Corrosion of Nuclear Power Plant Materials, NEA/CSNI/R(1994)26, Boulogne Billancourt, France (1995).
[92] OECD NUCLEAR ENERGY AGENCY, CODAP Project Topical Report on Basic Principles of Collecting and Evaluating Operating Experience Data on Metallic Passive Components, NEA/CSNI/R(2018)12, Boulogne Billancourt, France (2019).
[93] LYDELL, B., ESCRIG FORANO, D., RIZNIC, J., OECD Nuclear Energy Agency CODAP Database Project on passive component operating experience. An international collaboration in materials research, Nucl. Eng. Des. **380** (2021) 111280.
[94] VO, T.V., HEASLER, P.G., DOCTOR, S.R., SIMONEN, F.A., GORE, B.F., Estimates of rupture probabilities for nuclear power plant components: Expert judgment elicitation, Nucl. Technol. **96** (1991) 259–271.
[95] CHENG, W. C., et al., Review and categorization of existing studies on the estimation of probabilistic failure metrics

for reactor coolant pressure boundary piping and steam generator tubes in nuclear power plants, Prog. Nucl. Energy **118** (2020) 103105.

[96] BESUNER, P.M., TETELMAN, A.S., Probabilistic fracture mechanics, Nucl. Eng. Des. 43 (1977) 99–114.

[97] NUHI, M., ABU SER, T., AL TAMIMI, A.M., MODARRES, M., SEIBI, A., Reliability analysis for degradation effects of pitting corrosion in carbon steel pipes, Procedia Eng. **10** (2011) 1930–1935.

[98] NUCLEAR ENERGY AGENCY, EC JRC/OECD NEA Benchmark Study on Risk Informed In Service Inspection Methodologies (RISMET), NEA/CSNI/R(2010)13, Boulogne Billancourt, France (2010).

[99] BRICKSTAD, B. et al, WP 4, Review and Benchmarking of Structural Reliability Models and Associated Software, NURBIM Report DF, EURATOM, Brussels, Belgium (2004).

[100] HECKMAN, K., SAIFI, Q., Comparative analysis of deterministic and fracture mechanical assessment tools, Kerntech. **81** (2016) 484–497.

[101] QIAN, G., CHOU, H W., NIFFENEGGER, M., HUANG, C C., Probabilistic ageing and risk analysis tools for nuclear piping, Nucl. Eng. Des. **300** (2016) 541–551.

[102] YOSHIMURA, S., KANTO, Y., Probabilistic Fracture Mechanics for Risk Informed Activities. Fundamentals and Applications, Japan Welding Engineering Society, Tokyo, Japan (2017).

[103] U.S. NUCLEAR REGULATORY COMMISSION, Probabilistic Fracture Mechanics Regulatory Guide Update, No. ML19134A249 (2019).

[104] RAYNAUD, P., KIRK, M., BENSON, M., Important Aspects of Probabilistic Fracture Mechanics Analyses, TLR RES/DE/CIB 2018 1, U.S. Nuclear Regulatory Commission, Washington, DC (2018).

[105] FLEMING, K. et al, Treatment of Passive Component Reliability in Risk Informed Safety Margin Characterization, INL/EXT 10 20030, Idaho National Laboratory, Idaho Falls, ID (2010).

[106] UNWIN, S., et al., Physics Based SCC Reliability Model in a Cumulative Damage Framework, PNNL 20596, Pacific Northwest National Laboratory, Richland, WA (2011).

[107] VEERAMANY, A., PANDY, M., Reliability analysis of nuclear piping system using semi Markov process model, Ann. Nucl. Energy **38** (2011) 1133–1139.

[108] CHATTERJEE, K., MODARRES, M., A probabilistic physics of failure approach to prediction of steam generator tube rupture frequency, Nucl. Sci. Eng. **170** (2012) 136–150.

[109] ALDEMIR, T., Methodology Development for Passive Component Reliability Modeling in a Multi Physics Simulation Environment, Project Number 11 3030, Nuclear Energy University Programmes, U.S. Department of Energy, Idaho Falls, ID (2014).

[110] MAIO, F.D., COLLI, D., ZIO, E., TAO, L., TONG, J., A Multi State Physics Modeling Approach for the Reliability Assessment of Nuclear Power Plant Piping Systems, HAL Id: hal 01265883 (2016).

[111] SIMONEN, F.A., et al., Probabilistic Fracture Mechanics Evaluation of Selected Passive Components, PNNL 16625, Pacific Northwest National Laboratory, Richland, WA (2007).

[112] DUAN, X., WANG, M., KOZLUCK, M.J., Comparison of PRO LOCA 2009 and WinPRAISE 2007 for Estimation Rupture Probability of Dissimilar Metal Weld Susceptible to PWSCC, ICONE19 43921, Proc. Int. Conf. Nuclear Engineering, ASME, New York (2011).

[113] DUAN, X., WANG, M., KOZLUCK, M.J., Benchmarking PRAISE CANDU with nuclear risk based inspection methodology project fatigue cases, J. Pressure Vessel Technol. PVT 13 1217, (2015).

[114] SALTELLI, A. et al, Sensitivity Analysis. The Primer, John Wiley & Sons, Chichester (2008).

[115] ANDRES, T.H., Uncertainty Analysis Guide, AECL 12103, Atomic Energy of Canada Limited, Whiteshell Laboratories, Pinawa, Manitoba, Canada (2002).

[116] DEAN, V.F. (Ed.), Guide to the Expression of Uncertainties for the Evaluation of Critical Experiments, Rev. 4, Idaho National Laboratory, Idaho Falls, ID (2007).

[117] TRUE, D., VANOVER, D., Treatment of Parameter and Model Uncertainty for Probabilistic Risk Assessments, 1016737, EPRI, Palo Alto, CA (2008).

[118] ZIO, E., PEDRONI, N., Uncertainty Characterization in Risk Analysis for Decision Making Practice, Foundation for an Industrial Safety Culture, Toulouse, France (2012).

[119] DROUIN, M., et al., Guidance on the Treatment of Uncertainties Associated with PRA in Risk Informed Decision Making, NUREG 1855 Rev. 1, U.S. Nuclear Regulatory Commission, Washington, DC (2017).

[120] ERICKSON, M., Sources and Treatment of Uncertainties, xLPR Technical Report xLPR TR UNCERT Version 2.0 (2020).

[121] KLOOS, M., BERNER, N., SUSA: Software for Uncertainty and Sensitivity Analyses. Classical Methods, GRS 631, Gesellschaft für Anlagen und Reaktorsicherheit, Cologne, Germany (2021).

[122] U.S. NUCLEAR REGULATORY COMMISSION, An Approach for Determining the Technical Adequacy of

Probabilistic Risk Assessment Results for Risk Informed Activities, Regulatory Guide 1.200 Rev. 2, Washington, DC (2009).

[123] U.S. NUCLEAR REGULATORY COMMISSION, Plant Specific, Risk Informed Decision making for Inservice Inspections of Piping, Draft Regulatory Guide DG 1288, Rev. 1, Washington, DC (2020).

[124] MAIOLI, A., Newly Developed Method Requirements and Peer Review, PWROG 19027 NP, PWR Owners Group, Westinghouse Electric Company LLC, Cranberry Township, PA (2020).

[125] ZIO, E., PEDRONI, N., Risk Informed Decision Making Processes. An Overview, Foundation for an Industrial Safety Culture (FONCSI), Toulouse, France (2012).

[126] UK HEALTH & SAFETY EXECUTIVE, Probabilistic Methods: Uses and Abuses in Structural Integrity, Contract Research Report 398/2001, Sudbury, United Kingdom (2001).

[127] INTERNATIONAL ATOMIC ENERGY AGENCY, Safety of Nuclear Power Plants: Design, Specific Safety Requirements, No. SSR 2/1, Rev. 1, IAEA, Vienna (2016).

[128] AMERICAN SOCIETY OF CIVIL ENGINEERS, Minimum Design Loads and Associated Criteria for Buildings and Other Structures, ASCE/SEI 7 16, Reston, VA (2017).

[129] EUROPEAN COMMITTEE FOR STANDARDIZATION, Eurocode – Basis of Structural Design, E.N. 1990:2002, Brussels (2002).

[130] GRIESBACH, T., COFI, N., HARRIS, D., DEDHIA, D., ALLESHWARAM, A., Technical Basis for ASME Section XI Code Case on Flaw Tolerance of Cast Austenitic Stainless Steel Piping (MRP 362, Rev. 1), Report No. 3002007383, EPRI, Palo Alto, CA (2016).

[131] ASME, BPVC Section XI Rules for Inservice Inspection of Nuclear Power Plant Components, Division 1, Rules for Inspection and Testing of Components of Light Water Cooled Plants, BPVC XI 1 – 2021, ASME, New York, NY (2021).

[132] HARDIES, R., et al, xLPR Version 2.0 Technical Basis Document: Acceptance Criteria, U.S. Nuclear Regulatory Commission and EPRI, Washington, DC and Palo Alto, CA, NRC Public Document Room Accession No. ML16271A436 (2016).

[133] U.S. NUCLEAR REGULATORY COMMISSION, An Approach for Using Probabilistic Risk Assessment in Risk Informed Decisions on Plant Specific Changes to the Licensing Basis, Regulatory Guide 1.174, Rev. 2, Washington, DC (2011).

[134] U.S. NUCLEAR REGULATORY COMMISSION, Operability Determinations & Functionality Assessments for Resolution of Degraded or Non Conforming Conditions Adverse to Quality or Safety, NRC Inspection Manual, Part 9900: Technical Guidance, Washington, DC (2008).

[135] U.S. NUCLEAR REGULATORY COMMISSION, Risk Assessment of Operational Events Handbook, Vol. 1, Internal Events, Revision 2.02, SECY 99 007A, U.S. Nuclear Regulatory Commission, Washington, DC (1999).

[136] NUCLEAR STRUCTURAL INTEGRITY PROBABILISTIC WORKING GROUP, Nuclear Structural Integrity Probabilistic Working Principles, UK Forum for Engineering Structural Integrity (FESI), Fulwood, Preston, United Kingdom (2019).

[137] KURISAKA, K., NAKAI, R., ASAYAMA, T., TAKAYA, T., Development of system based code (1) reliability target derivation of structures and components, J. Power Energy Syst. **5** 1 (2011) 19–32.

[138] ERICKSONKIRK, M.T., DICKSON, T.L., Recommended Screening Limits for Pressurized Thermal Shock (PTS), NUREG 1874, U.S. Nuclear Regulatory Commission, Washington, DC (2007).

[139] RUDLAND, D., HARRINGTON, C., xLPR Pilot Study Report, NUREG 2110, U.S. Nuclear Regulatory Commission, Washington, DC (2012).

[140] RIGDON, S., BASU, A., Statistical Methods for the Reliability of Repairable Systems, John Wiley and Sons Inc., New York, (2000).

[141] PANDEY, M., Probabilistic Assessment: Principles and Computational Methods, Research Project R706.1 Final Report, Canadian Nuclear Safety Commission, Ottawa, Canada (2020).

[142] LAXMAN, S., BLAIR, C., "Regulatory Perspective for the Definition of Probabilistic Acceptance Criteria for CANDU Pressure Tubes", (Proc. ASME 2016 Pressure Vessels and Piping Conf.), PVP2016 63655, ASME, New York (2016).

[143] SCARTH, D., GUTKIN, L., "Acceptance Criteria for Probabilistic Fracture Protection Evaluations of CANDU Zr Nb Pressure Tubes", (Proc. ASME 2018 Pressure Vessels and Piping Conf.), PVP2018 85086, ASME, New York (2018).

[144] SALLABERRY, C.J., KURTH, R., Sensitivity Studies and Analyses Involving the Extremely Low Probability of Rupture Code, TLR/RES/DE/CIB 2021 11, U.S. Nuclear Regulatory Commission, Washington, DC (2021).

[145] SOMASUNDARAN, D., DEDHIA, D., SHIM, D.J., HARRINGTON, C., "Sensitivity Analysis Methodology for Probabilistic Fracture Mechanics Output", (Trans. SMiRT 25, Charlotte, NC, August 4–9, 2019).

[146] HECKMANN, K., et al., Comparison of sensitivity measures in probabilistic fracture mechanics, Int. J. Press. Vessels Piping **192** (2021) 104388.

[147] BEAL, J., SAKURAHARA, T., REIHANI, S., KEE, E., MOHAGHEGH, Z., "An Algorithm for Risk Informed Analysis of Advanced Reactors with a Case Study of Pipe Failure Rate Estimation", (30th European Safety and Reliability Conf. and 15th Probabilistic Safety Assessment and Management Conf.) ESREL 2020 & PSAM 15, Venice, Italy (2020).

[148] FÉRON, D., STAEHLE, R.W. (Eds), Stress Corrosion Cracking of Nickel base Alloys in Water Cooled Nuclear Reactors. The Coriou Effect, Elsevier S&T Books, Duxford, UK (2016).

[149] FÉRON, D., GUERRE, C., MARTIN, F., Historical review of Alloy 600 stress corrosion cracking: From the "Coriou effect" to the quantitative micro nano approach, Corros. J. **78** (2019) 267–273.

[150] MOSS, T., WAS, G.S., "Factor of Improvement in Resistance of in Stress Corrosion Cracking Initiation of Alloy 690 over Alloy 600", Proc. 17th Int. Conf. on Environmental Degradation of Materials in Nuclear Power Systems – Water Reactors, August 9–13, 2015, Ottawa, Ontario, Canada (2017).

[151] XU, H., et al., Materials Reliability Program (MRP). Resistance to Primary Water Stress Corrosion Cracking of Alloys 690, 52 and 152 in Pressurized Water Reactors (MRP 111), 1009801, EPRI, Palo Alto, CA (2004).

[152] HICKLING, J., Materials Reliability Program: Resistance of Alloys 690, 52 and 152 to Primary Water Stress Corrosion Cracking (MRP 237, Rev. 1). Summary of Findings from Completed and Ongoing Test Programs since 2004, 1018130, EPRI, Palo Alto, CA (2008).

[153] MOSS, T., KUANG, W., WAS, G.S., Stress corrosion cracking initiation in Alloy 690 in high temperature water, Curr. Opin. Solid State Mater. Sci. **22** (2018) 16–25.

[154] JENKS, A., WHITE, G., BURKARDT, M., YOUNG, G., Materials Reliability Program: Recommended Factors of Improvement for Evaluating Primary Water Stress Corrosion Cracking (PWSCC) Growth Rates of Thick Walled Alloy 690 Materials and Alloy 52, 152 and Variants Welds (MRP 386), 3002010756, EPRI, Palo Alto, CA (2018).

[155] JENKS, A., WHITE, G., BURKARDT, M., Materials Reliability Program: Crack Growth Rates for Evaluating Primary Water Stress Corrosion Cracking (PWSCC) of Thick Wall Alloy 600 Materials and Alloy 82, 182, and 132 Welds (MRP 420, Revision 1), 3002014244, EPRI, Palo Alto, CA (2018).

[156] SMITH, C.L., Calculating conditional core damage probabilities for nuclear power plant operations, Reliab. Eng. Syst. Saf. **59** (1998) 299–307.

[157] BABCOCK & WILCOX COMPANY, 177 Fuel Assembly Owner's Group Safe End Task Forde Report on Generic Investigation of High Pressure Safety Injection /MU Nozzle Component Cracking, Report No. 77 1140611 00, Lynchburg, VA (1983).

[158] BWR VESSEL & INTERNALS PROJECT (BWRVIP), White Paper on Suggested Content for PFM Submittals to the NRC, EPRI, Palo Alto, CA (2019). (NRC ADAMS Accession No. ML19241A545).

[159] U.S. NUCLEAR REGULATORY COMMISSION, Plant Specific, Risk Informed Decision Making for In Service Inspections of Piping, Regulatory Guide 1.178 Rev. 2, Washington, DC (2021).

[160] INTERNATIONAL ATOMIC ENERGY AGENCY, Use of a Graded Approach in the Application of the Management System Requirements for Facilities and Activities, IAEA TECDOC 1740, IAEA, Vienna (2014).

[161] U.S. NUCLEAR REGULATORY COMMISSION, PFM Submittal Guidance, Draft Regulatory Guide DG 1382, Washington, DC (2021).

[162] U.S. NUCLEAR REGULATORY COMMISSION, Technical Basis for the Use of Probabilistic Fracture Mechanics in Regulatory Applications, NUREG/CR 7278, Washington, DC (2022).

[163] FLEMING, K., LYDELL, B., Guidelines for Performance of Internal Flooding Probabilistic Risk Assessment, Product ID 1019194, Palo Alto, CA (2009).

[164] ELECTRIC POWER RESEARCH INSTITUTE, FRANX Software Manual, Product ID 3002010659, EPRI, Palo Alto, CA (2017).

[165] FLEMING, K., LYDELL, B., Pipe Rupture Frequencies for Internal Flooding Probabilistic Risk Assessment, Report No. 3002000079, EPRI, Palo Alto, CA (2013).

[166] SWALING, V.H., R Book. Reliability Data Handbook for Piping Components in Nordic Nuclear Power Plants, Phase II: ASME Code Class 1 & 2, Summary Report, NPSAG Report 04 007:01, Nordic PSA Group, Stockholm, Sweden (2010).

[167] SWALING, V.H., OLSSON, A., Reliability Data Handbook for Piping Components in Nordic Nuclear Power Plants: R Book Phase II, SSM 2011:06, Swedish Radiation Safety Authority, Stockholm, Sweden (2011).

[168] ASOMOLOV, V.G., GUSEV, I.N., KAZANSKI, V.R., POVAROV, V.P., STATSURA, D.B., New generation first of the kind unit – VVER 1200 design features, Nucl. Energy Technol. **3** (2017) 260–269.

[169] FENNERN, L., Design evolution of BWRs: Dresden to generation III+, Prog. Nucl. Energy **102** (2018) 38–57.

[170] CUMMINS, W.E., MATZIE, R., Design evolution of PWRs: Shippingport to generation III+, Prog. Nucl. Energy **102** (2018) 9–37.

[171] FERNÁNDEZ ARIAS, VERGARA, D., OROSA, J.A., A global review of PWR nuclear power plants, Appl. Sci. **10** (2020) 4434.

[172] MEHTA, H.S., A Review of NUREG/CR 5750 Intergranular SCC Improvement Factor and Probability of Rupture Given a Through Wall Crack, GE NE A41 00110 00 1 Rev. 0, GE Nuclear Energy, San Jose, CA (2002).

[173] LYDELL, B., Estimation of HDPE Pipe Failure Rates & Rupture Frequencies. A Feasibility Study, SPI 2020 R02, A Technical Report Prepared for the Electric Power Research Institute, Sigma Phase Inc., Vero Beach, FL (2020).

[174] NIE, J., et al., Identification and Assessment of Recent Aging Related Degradation Occurrences in U.S. Nuclear Power Plants, BNL 81741 2008, Brookhaven National Laboratory, Upton, NY (2008).

[175] WOLFORD, A.J., ATWOOD, C.L., ROESNER, W.S., WEIDENHAMER, G.H., Aging Data Analysis and Risk Assessment. Development and Demonstration Study, NUREG/CR 5378, U.S. Nuclear Regulatory Commission, Washington, DC (1992).

[176] SANZO, D., et al., Survey and Evaluation of Aging Risk Assessment Methods and Applications, NUREG/CR 6157, U.S. Nuclear Regulatory Commission, Washington, DC (1994).

[177] SMITH, C.L., SHAH, V.N., KAO, T., APOSTOLAKIS, Incorporating Aging Effects into Probabilistic Risk Assessment – A Feasibility Study Utilizing Reliability Physics Models, NUREG/CR 5632, U.S. Nuclear Regulatory Commission, Washington, DC (2001).

[178] RODIONOV, A., ATWOOD, C.L., KIRCHSTEIGER, C., PATRIK, M., Demonstration of statistical approaches to identify component's aging by operational data analysis – A case study for the aging PSA network, Reliab. Eng. Syst. Saf. **93** (2008) 1534–1542.

[179] RODIONOV, A., KELLY, D., UWE KLÜGEL, J., Guidelines for Analysis of Data Related to Ageing of Nuclear Power Plant Components and Systems, EUR 23954 EN, European Commission, Joint Research Centre, Petten, The Netherlands (2009).

[180] GETMAN, A.F., ARKADOV, G.V., RODIONOV, A.N., Probabilistic Safety Assessment for Optimum Nuclear Power Plant Life Management (PLiM): Theory and Application of Reliability Analysis Methods for Major Power Plant Components, Elsevier Science & Technology, Cambridge, MA (2012).

[181] OECD NUCLEAR ENERGY AGENCY, A Review of Operating Experience Involving Passive Component Material Degradation in Periods of Extended / Long Term Operation, NEA Publication 7614, Boulogne Billancourt, France (2022).

[182] LYDELL, B., FLEMING, K., ROY, J. F., "Analysis of Possible Aging Trends in the Estimation of Piping System Failure Rates for Internal Flooding PRA", (Proc. PSAM 14: Probabilistic Safety Assessment and Management, Los Angeles, CA, September 16–21, 2018), Paper #367.

[183] MOROZOV, V., TOKMACHEV, G., LUTSUK, R., KOPYLOV, S., Full scope probabilistic safety assessment of Balakovo Unit 1 in Russia, J. Nucl. Eng. Radiat. Sci. 3 (2017) 011016 1.

CONTENTS OF THE ANNEXES

The on-line supplementary files for this publication can be found on the publication's web page at https://www.iaea.org/publications.

GLOSSARY

ageing. IAEA Safety Glossary[1]:

"General *process* in which characteristics of a *structure, system or component* gradually change with time or use.

"Although the term *ageing* is defined in a neutral sense — the changes involved in ageing may have no effect on *protection* or *safety*, or could even have a beneficial effect — it is most commonly used with a connotation of changes that are (or could be) detrimental to *protection and safety* (i.e. as a synonym of *ageing degradation*).

"***non-physical ageing.*** The *process* of becoming out of date (i.e. obsolete) owing to the evolution of knowledge and technology and associated changes in codes and standards.

- — Examples of *non-physical ageing* effects include the lack of an effective *containment* or *emergency* core cooling system, the lack of *safety design* features (such as *diversity*, separation or *redundancy*), the unavailability of qualified spare parts for old equipment, incompatibility between old and new equipment, and outdated *procedures* or documentation (e.g. which thus do not comply with current regulations).
- — Strictly, this is not always *ageing* as defined above, because it is sometimes not due to changes in the *structure, system or component* itself. Nevertheless, the effects on *protection and safety*, and the solutions that need to be adopted, are often very similar to those for *physical ageing*.
- — The term technological obsolescence is also used.

physical ageing. Ageing of structures, systems and components due to physical, chemical and/or biological processes (ageing mechanisms).

- — Examples of *ageing* mechanisms include wear, thermal or *radiation* embrittlement, corrosion and microbiological fouling."

ageing degradation. IAEA Safety Glossary, 2016 Revision, June 2016:

"*Ageing* effects that could impair the ability of a *structure, system or component* to function within its *acceptance criteria*.

- — Examples include reduction in diameter due to wear of a rotating shaft, loss in material toughness due to *radiation* embrittlement or thermal *ageing*, and cracking of a material due to fatigue or stress corrosion cracking."

ageing management. IAEA Safety Glossary, 2016 Revision, June 2016:

"Engineering, operations and maintenance actions to control within acceptable limits the ageing degradation of structures, systems and components.

---

[1] INTERNATIONAL ATOMIC ENERGY AGENCY, IAEA Safety Glossary: 2018 Edition, Non-serial Publications, IAEA, Vienna (2019).

— Examples of engineering actions include *design*, *qualification* and *failure analysis*. Examples of *operations* actions include *surveillance*, carrying out operating *procedures* within specified *limits* and performing environmental measurements.

"**life management (or lifetime management).** The integration of *ageing management* with economic planning: (1) to optimize the *operation*, *maintenance* and *service life* of *structures, systems and components*; (2) to maintain an acceptable level of *safety* and performance; and (3) to improve economic performance over the *service life* of the *facility*."

ageing PSA. A PSA model which models ageing (or time dependent) effects in component failure rates. A 'classic' PSA uses the assumption that component failure rates are constant.

aleatory uncertainty. This form of uncertainty is associated with the randomness of events such as component failures, initiating events and hazard events. This is addressed by modelling these events using probability models, for example, these are typically Poisson and binomial distributions for initiating events and component failures. These probability models represent initiating event frequencies and component failure probabilities or failure rates.

ASA/ANSI B31.1 piping. The piping systems in the early US nuclear plants were designed in accordance with the requirements of American Standards Association (ASA) B31.1-1955. Section 1 of B31.1-1955 was written for Power Piping, and encompassed the "... minimum requirements for design, manufacture, test, and installation of power piping systems, as defined for steam generating plants, central heating plants, and industrial plants." B31.1-1955 included specific requirements for pipe wall thickness and allowable stresses.

ASME Section III piping. Division 1 of ASME Boiler and Pressure Vessel Code Section III contains requirements for piping classified as ASME Class 1, Class 2 and Class 3. ASME Section III does not delineate the criteria for classifying piping into Class 1, Class 2 or Class 3; it specifies the requirements for design, materials, fabrication, installation, examination, testing, inspection, certification and stamping of piping systems after they have been classified Class 1, Class 2 or Class 3 based upon the applicable design criteria and Regulatory Guide 1.26, Quality Group Classifications and Standards for Water-Steam, and Radio-Waste-Containing Components of NPPs. Subsections NB, NC and ND of ASME III specify the construction requirements for Class 1, Class 2 and Class 3 components, including piping, respectively. Subsection NF contains construction requirements for component supports, and a newly added subsection NH contains requirements for Class 1 Components in Elevated-Temperature Service. Subsection NCA, which is common to Divisions 1 and 2, specifies general requirements for all components within the scope of ASME Section III.

ASME Section XI. The ASME Code Section XI (Rules for Inservice Inspection of Nuclear Power Plant Components) provides standards for the examination, in-service testing and inspection, and repair and replacement of NPP components, pressure vessels and piping. The code details if a flaw found within a component is acceptable for continued operation, or if the component instead requires repair or replacement. The first edition of ASME Section XI was published on 1 January 1971.

ASME Section XI, DIVISION 2. Issued in July 2019, Division 2 provides the requirements for the creation of an RIM programme for advanced nuclear reactor designs. An RIM programme addresses the entire life cycle for all types of NPPs; it requires a combination of monitoring, examination, tests, operation and maintenance requirements that ensures each SSC meets plant risk and reliability goals that are selected for the RIM programme.

augmented in-service inspection programme. An inspection programme that has been developed to address observed material degradation, and an inspection is targeted at locations where the most severe degradation is expected.

austenitic alloy steel. High alloy steels with the main alloying elements being chromium (Cr) and nickel (Ni). Some high alloy steels include niobium (Nb) to improve welding properties, or titanium (Ti) to prevent intergranular corrosion and weld decay.

balance of plant. It consists of the remaining systems, components, and structures that comprise a complete NPP and are not included in the nuclear steam supply system.

Bayesian reliability analysis. In the Bayesian approach, an SME develops a well-informed estimate of the probability of failure distribution; the prior state of knowledge. This probability distribution is then updated as more information is collected about the structural integrity of a certain piping system component.

buttering. This is the adding process of material in welding. Weld metal is deposited on one or more surfaces to provide a metallurgically compatible weld metal for the subsequent completion of the weld.

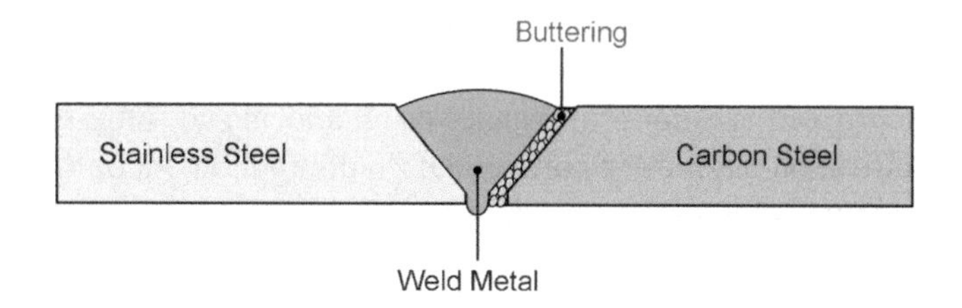

calibration. Calibration is an analytical process that is used to adjust a set of parameters associated with a computational science and engineering code so that the model agreement is maximized with respect to a set of experimental data or operating experience data.

cavitation. When the static pressure of a liquid reduces to below its vapour pressure, small vapour-filled cavities are formed in the liquid. This is called cavitation. The cavities are also called bubbles or voids. When imposed to higher pressure, the cavities collapse in, thus generating shock waves (they are stronger closer to the imploding cavity and weaken as they propagate away from it). Therefore, cavitation damage of significance in engineering systems is created by collapsing cavities near the metal surface causing its fatigue due to generated cyclic stress. Most often, cavitation causes most of its damage by vibration (e.g. cracked welds, broken instrument lines, loosened flanges). The erosion caused by cavitation also generates particles that contaminate the process fluid.

code repair. A US term, the definitions of and requirement for a 'Code Repair' are defined in ASME Section XI, Article IWA-4000, Repair/Replacement Activities.

cold working. Cold working refers to the process of strengthening metal by changing its shape without the use of heat. Subjecting the metal to this mechanical stress changes the mechanical properties and the environmentally assisted crack growth rate. The crack growth rate increases rapidly with the increase of the level of cold working.

component boundary. This defines the physical boundary of a component required for system operation. A component boundary definition should be consistent with the parameter database supporting

PSA model quantification. For piping components, the component boundary is established through degradation mechanism evaluations (see below).

Coriou effect. In 1959, the French materials scientist Henri Coriou and his co-workers published the results of laboratory tests which indicated that Alloy 600 in a high temperature (350°C) pure water environment is susceptible to primary water SCC. The so-called 'Coriou effect' was met with skepticism until the early 1980s when a growing number of primary water SCC failures were reported.

crevice corrosion. Crevice corrosion occurs in a wetted or buried environment when a crevice or area of stagnant or low flow exists that allows a corrosive environment to develop in a component. It occurs most frequently in joints and connections, or points of contact between metals and non-metals, such as gasket surfaces, lap joints and under bolt heads. Carbon steel, cast iron, low alloy steels, stainless steel, copper and nickel base alloys are all susceptible to crevice corrosion. Steel can be subject to crevice corrosion in some cases after lining/cladding degradation.

critical crack length. The length of a crack (either axial or circumferential) for which the crack will propagate unstably for a given set of loading conditions.

dealloying (selective leaching). 'Dealloying' or 'selective leaching' refers to the selective removal of one element from an alloy by corrosion processes. A common example is the dezincification of unstabilized brass, whereby a weakened, porous copper structure is produced. The selective removal of zinc can proceed in a uniform manner or on a localized (plug-type) scale. It is difficult to rationalize dezincification in terms of preferential Zn dissolution out of the brass lattice structure. Rather, it is believed that brass dissolves with Zn remaining in solution and Cu replating out of the solution. Graphitic corrosion of grey cast iron, whereby a brittle graphite skeleton remains following preferential iron dissolution is a further example of selective leaching. During cast iron graphitic corrosion, the porous graphite network that makes up 4–5% of the total mass of the alloy is impregnated with insoluble corrosion products. As a result, the cast iron retains its appearance and shape but is weaker structurally. Testing and identification of graphitic corrosion is accomplished by scraping through the surface with a knife to reveal the crumbling of the iron beneath.

degradation mechanism. Phenomena or processes that attack (wear, erode, crack, etc.) the pressure-retaining material over time and might result in a reduction of component pressure boundary integrity. It should be noted that damage mechanisms and degradation mechanisms could interact to cause major, catastrophic passive component failures.

delayed hydride cracking (DHC). A subcritical crack growth mechanism occurring in zirconium alloys as well as other hydride-forming materials that requires the formation of brittle hydride phases at the tip of a crack and subsequent failure of that hydride resulting in crack extension. Hydrogen in solution in the zirconium alloy is transported to the crack tip by diffusion processes where it precipitates as a hydride phase. When the precipitate attains a critical condition, related to its size and the applied stress intensity factor, K_I, fracture ensues and the crack extends through the brittle hydride and arrests in the matrix. Each step of crack propagation results in crack extension by a distance approximately the length of the hydride.

design certification PSA. In the US a design certification (DC) is achieved through a regulatory rulemaking process, which addresses the various safety issues associated with the proposed NPP design, independent of a specific site. A DC application contains a final safety analysis report (FSAR) that is supported by a design specific PSA.

displacement-controlled stresses. Stresses that result from the application of displacements, such as those due to thermal expansion or seismic anchor motion.

double-ended guillotine break. A condition for which a circumferential through-wall crack propagates around the entire circumference of the pipe such that the cracked pipe section severs into two pieces and the two ends are displaced relative to their pipe axes to allow for full flow from each end.

ductile crack growth. With ductile fracture a crack moves slowly and is accompanied by a large amount of plastic deformation around the crack tip. A ductile crack will usually not propagate unless an increased stress is applied and generally cease propagating when loading is removed.

elementary effects method. Also referred to as Morris' elementary effects method. The elementary effects method is applied to identify non-influential inputs for a computationally elaborate mathematical model or for a model with a large number of inputs, where the costs of estimating other sensitivity analysis measures such as the variance based measures is not affordable. The elementary effects method provides qualitative sensitivity analysis measures (i.e. measures which allow the identification of non-influential inputs or which allow one to rank the input factors in order of importance, but do not quantify exactly the relative importance of the inputs).

enhanced visual examination. The enhanced visual examination -1 method is intended for the visual examination of surface breaking flaws. Any visual inspection for cracking requires a reasonable expectation that the flaw length and crack mouth opening displacement meet the resolution requirements of the observation technique. The enhanced visual examination -1 specification augments the VT-i requirements to provide more rigorous inspection standards for SCC.

environmentally assisted cracking. A localized deformation process accelerated by local corrosion in addition to mechanical stresses or strains. The cracking of structural materials in NPPs may proceed along grain boundaries (i.e. intergranular SCC), underlining the role of dissolution, or through the grains (i.e. transgranular SCC), underlining the role of mechanical loading.

epistemic uncertainty. The uncertainty that comes from a lack of knowledge. This lack of knowledge comes from many sources. Inadequate understanding of the underlying processes, incomplete knowledge of the phenomena, or imprecise evaluation of the related characteristics. For rare events the failure parameters may be derived from extrapolation using statistical models and sparse historical data.

equivalent break size. The calculated size of a hole in a pipe given certain through-wall flow rates and for a given pressure.

erosion cavitation. Erosion cavitation is the process when a material surface deteriorates due to the creation of vapour or gas pockets inside the flow of liquid in thus causing surface material loss. Once cavitation is present, the erosion is caused by the bombardment of vapour bubbles on the material surface. When these bubbles strike the surface, they collapse, or implode. Although a single bubble imploding does not carry much force, over time, the small damage caused by each bubble accumulates. The repeated impact of these implosions results in the formation of pits. Also, like erosion, the presence of chemical corrosion enhances the damage and rate of material removal. Erosion-cavitation has been observed in PWR stainless steel decay heat removal and charging system piping.

erosion-corrosion. Erosion is the destruction of metals by the abrasive action of moving fluids, usually accelerated by the presence of solid particles or matter in suspension. When corrosion occurs simultaneously, the term 'erosion-corrosion' is used. This term applies to moderate energy carbon steel piping (e.g. raw water piping).

factor of improvement. Also referred to as 'relative improvement factor', the factor of improvement is an estimate of the crack growth rate in one type of material relative to another material type. It is an indication of the estimated improvement in material performance as a function of material chemical composition, mechanical properties, local stresses, degradation mitigation techniques, etc. The factor of improvement can be determined on the basis of testing in laboratory environments, analytical work (e.g. statistical analysis of test data), expert elicitation and or operating experience data.

failure. A condition for which a component or system is no longer capable of performing its design function. Depending on the context, failure can be defined as either the condition for which the piping system is no longer capable of maintaining internal pressure, or when the pipe experiences a double-ended guillotine break. National codes and standards (e.g. the ASME Boiler and Pressure Vessel Code, Section XI, Rules for Inservice Inspection of Nuclear Power Plant Components) give specific details on what constitutes a failure.

ferrite content. Ferrite is the ferromagnetic, body centred cubic, microstructural constituent of variable chemical composition in iron-chromium-nickel alloys. Austenitic stainless steels are essentially free of ferrite, which is magnetic. Cast products of these alloys typically have some ferrite present. These alloys also form some ferrite when they are cold worked or work hardened. In both cases the products will show a magnetic tendency. Ferrite can be detrimental to corrosion resistance in some environments. There are also applications where magnetic characteristics interfere with performance of the end product. The ferrite content of the cast alloy can be controlled through alloy composition. Carbon, nitrogen, nickel and manganese are strong austenite formers and increasing their content in the alloy will reduce the tendency for ferrite formation. There are several different methods of predicting the ferrite content but one of the more common is the DeLong diagram. Ferrite reduces the steel's tendency for solidification cracking during cooling. It is not uncommon for 304 castings (CF8) to contain 8–20% ferrite. The cast ingot composition of wrought 304 stainless is also balanced to have 1–6% ferrite since this reduces the chance of cracking during forging or hot working.

flashing. Flashing occurs when a high pressure liquid flows through a valve or an orifice to a region of greatly reduced pressure. If the pressure drops below the vapour pressure, some of the liquid will be spontaneously converted to steam. The downstream velocity will be greatly increased due to a much lower average density of the two-phase mixture. The impact of the high velocity liquid on piping or components creates flashing damage.

flaw. An imperfection or unintentional discontinuity that is detectable by non-destructive examination.

flaw aspect ratio. Ratio of the length of the deepest crack to the depth of the deepest crack.

flow accelerated corrosion. Flow accelerated corrosion is a process whereby the normally protective oxide layer on carbon or low alloy steel dissolves into a stream of flowing water or water-steam mixture. It can occur in both single-phase and two-phase regions. The cause of flow accelerated corrosion is a specific set of water chemistry conditions (e.g. pH, level of dissolved oxygen), and there is no mechanical contribution to the dissolution of the normally protective iron oxide (magnetite) layer on the inside pipe wall.

flow induced vibration. The term 'flow induced vibration' is used to describe outside diameter pipe wear (or wall loss) caused by the interaction of flow induced vibration and reflective metal insulation

full structural weld overlay (FSWOL). A structural reinforcement and SCC mitigation technique through application of an SCC resistant material layer around the entire circumference of the treated weldment. The minimum acceptable FSWOL thickness is 1/3 the original pipe wall thickness. The minimum length is $0.75\sqrt{(R \times t)}$ on either side of the dissimilar metal weld to be treated, where R is the outer radius of the item and t is the nominal thickness of the item.

general corrosion. An approximately uniform wastage of a surface of a component, through chemical or electrochemical action, free of deep pits or cracks.

heat affected zone. The heat affected zone is an area between the weld or cut and the base metal, and is a non-melted metal's area that was exposed to high temperatures and has therefore undergone changes in its properties. It can vary in its size and severity, dependent on the materials involved, the intensity and concentration of heat, and the process employed. Welding with high heat input (i.e. fast heating) has faster cooler rates compared to welding with low heat input (i.e. slow heating) and thus, has smaller heat affected zones. Conversely, welding with low heat input results in a larger heat affected zone. As the speed of the welding decreases, the size of a heat affected zone increases. Weld geometry also plays a role in the size of the heat affected zone. The heat affected zone's problems can be mitigated by performing a pre- and/or post-weld heat treatment. In the high temperature cutting processes, the depth of the heat affected zone is associated with the cutting process itself, cutting speed, material properties and material thickness. Similar to welding, cutting metals at high temperatures and slow speeds tend to lead to large heat affected zones. Cutting metal at high speeds tends to reduce the width of the heat affected zone because the zone experiences sufficient heat for a long enough time, and the layer undergoes microstructure and property changes that differ from the parent metal. These changes are usually undesirable and ultimately serve as the weakest part of the component. For example, the microstructural changes can lead to high residual stresses, reduced material strength, increased brittleness and decreased resistance to corrosion and/or cracking. As a result, many failures occur in the heat affected zone.

high energy piping. A piping system for which the maximum operating temperature exceeds 94 °C or the maximum operating pressure exceeds 1.9 MPa.

hydrogen induced cracking. Stepwise internal cracks that connect adjacent hydrogen blisters on different planes in the metal, or to the metal surface. An externally applied stress is not needed for the formation of hydrogen induced cracking. In steels, the development of internal cracks (sometimes referred to as blister cracks) tends to link with other cracks by a transgranular plastic shear mechanism because of internal pressure resulting from the accumulation of hydrogen. The link-up of these cracks on different planes in steels has been referred to as stepwise cracking to characterize the nature of the crack appearance. Hydrogen induced cracking is commonly found in steels with: (a) high impurity levels that have a high density of large planar inclusions, and/or (b) regions of anomalous microstructure produced by segregation of impurity and alloying elements in the steel.

hydrostatic pressure test. A pressure test conducted during a plant or system shutdown at a pressure above nominal operating pressure or system pressure for which overpressure protection is provided.

induction heat stress improvement. The induction heat stress improvement process changes the normally tensile stress on the inside diameter surface of weld heat affected zones to compression. This process involves induction heating of the outer pipe surface of completed girth welds, while simultaneously water cooling the inside with flowing water. Thermal expansion caused by the heating coil yields the outside surface in compression, while the cool inside surface yields in tension.

After cooldown, contraction of the outside diameter causes the direction of stress to reverse, leaving the inside diameter in compression and the outside diameter in tension.

latent failure. A degraded material condition that may lie dormant for a long period before leading to a visible flaw (e.g. through-wall crack, active leakage).

leak before break (LBB). Generally referred to as a methodology whereby it is shown that a crack can be detected by leakage under normal operating conditions and that that crack would be stable at normal plus safe shutdown earthquake loads. Sometimes also referred to as a condition whereby a surface crack breaks through the pipe thickness and remains stable even if the break-through occurs at emergency or faulted loads. The LBB principle does not apply to systems that are susceptible to fatigue cracking, SCC or water hammer.

limiting condition for operation (LCO). According to the technical specifications, an LCO is the lowest functional capability or performance level of a piece of equipment required for safe operation of a nuclear plant. When an LCO cannot be met, the reactor is to be shut down or the licensee follows any remedial action permitted by the technical specifications until the condition can be met.

liquid droplet impingement. The liquid droplet impingement is caused by the impact of high velocity droplets or liquid jets. Normally, liquid droplet impingement occurs when a two-phase stream experiences a high pressure drop (e.g. across an orifice on a line to the condenser). When this occurs, there is an acceleration of both phases with the liquid velocity increasing to the point that, if the liquid strikes a metallic surface, damage to the surface will occur. The main distinction between flashing and liquid droplet impingement is that in flashing the fluid is of lower quality (mostly liquid with some steam), and with liquid droplet impingement, the fluid is of higher quality (mostly steam with some liquid).

loss of coolant accident. Those postulated accidents that result in a loss of reactor coolant at a rate in excess of the capability of the reactor make-up system from breaks in the reactor coolant pressure boundary, up to and including a break equivalent in size to the double-ended rupture of the largest pipe of the RCS.

mechanical stress improvement process (MSIP). A patented process that was invented, developed and first used in 1986 for mitigating SCC in nuclear plant weldments. MSIP works by using a hydraulically operated clamp which contracts the pipe on one side of the weldment. A typical tool design consists of a specially designed hydraulic box press for bringing the clamp halves together. By contracting the pipe on one side of the weldment, the residual tensile stresses are replaced with compressive stresses.

moderate energy piping. A piping system for which the maximum operating temperature is less than 94°C or the maximum operating pressure is less than 1.9 MPa.

nominal pipe size. A North American set of standard sizes for pipes used for high or low pressures and temperatures. 'Nominal' refers to pipe in non-specific terms and identifies the diameter of the hole with a non-dimensional number (for example, a 2 inch nominal steel pipe consists of many varieties of steel pipe with the only criterion being a 2.375 inch OD). Specific pipe is identified by pipe diameter and another non-dimensional number for wall thickness referred to as the Schedule (Sched. or Sch., for example, –2-inch diameter pipe, Schedule 40). The European and international designation equivalent to nominal pipe size is DN (diamètre nominal/nominal diameter/Durchmesser nach Norm), in which sizes are measured in millimetres.

probability of detection. The probability that a flaw of a certain size will be detected; it is conditional on factors such as wall thickness, non-destructive examination personnel qualifications and flaw orientation.

radiographic examination. A non-destructive testing method of inspecting materials for hidden flaws by using the ability of short wavelength electromagnetic radiation (high energy photons) to penetrate various materials.

reliability and integrity management. Those aspects of the plant design and operational phase that are applied to provide an appropriate level of reliability of SSCs and a continuing assurance over the life of the plant that such reliability is maintained. The most recent edition of the ASME Boiler and Pressure Vessel Code Section XI was issued in 2019. Division 2 of the 2019 edition provides the requirements for the creation of the RIM programme for advanced nuclear reactor designs. The RIM programme addresses the entire life cycle for all types of NPPs. It requires a combination of monitoring, examination, tests, operation and maintenance requirements that ensures each SSC meets plant risk and reliability goals that are selected for the RIM programme.

repair weld. A weld subjected to high residual stresses as a result of extensive repairs (e.g. during original construction). Such regions may be subjected to premature SCC.

residual stresses. Those stresses that remain in an object (in particular, in a welded component) even in the absence of external loading or thermal gradients. In some cases, residual stresses result in significant plastic deformation, leading to warping and distortion of an object. In others, they affect susceptibility to fracture and fatigue.

risk-informed in-service inspection. Risk-informed ISI methodologies are applied to determine the risk significance and failure potential of a piping component. This enables the targeting in-service examination resources to locations that are truly risk significant, providing the ability to capture or minimize risk and thereby improving plant reliability while keeping radiation doses to workers as low as reasonably achievable. The failure potential is determined through degradation mechanism analysis. The risk significance is determined through pipe failure frequency assessment.

round-robin test. An analysis or experiment performed independently several times and by different teams. This can involve multiple teams performing analyses using a common set of data with the use of a variety of methods. A round-robin test may be conducted to determine the reproducibility of an analysis; from input parameters to results interpretation.

selective leaching. Also referred to as dealloying, demetalification, parting and selective corrosion, it is a corrosion type in some solid solution alloys, when in suitable conditions a component of the alloys is preferentially leached from the material. The less noble metal is removed from the alloy by a microscopic scale galvanic corrosion mechanism. The most susceptible alloys are those containing metals with high distance between each other in the galvanic series (e.g. copper and zinc in brass).

***s-n* curve.** A plot of the magnitude of an alternating stress versus the number of cycles to failure for a given material. Typically, both the stress and number of cycles are displayed on logarithmic scales. *S-N* curves were developed by the German scientist, August Wöhler, during the investigation of an 1842 train crash in Versailles, France. In this crash, the axle of the train locomotive failed under the repeated 'low level' cyclic stress of everyday usage on the railroad. While investigating, Wöhler discovered that cracks formed and slowly grew on an axle surface. The cracks, after reaching a critical size, would suddenly propagate and the axle would fail. The level of these loads was less than the ultimate strength and/or yield strength of the material used to manufacture the axle.

socket weld. A pipe attachment detail in which a pipe is inserted into a recessed area of a pipe (e.g. elbow), valve or flange; see figure below[2]. Socket welds are mainly used for small pipe diameters; generally for piping whose nominal diameter is DN50 or smaller.

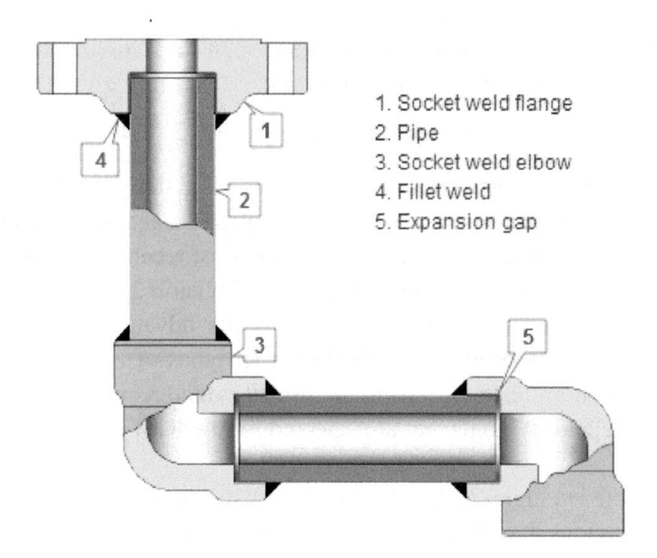

solid particle erosion. Solid particle erosion represents damage that is caused by particles transported by fluid stream rather than by liquid water or collapsing bubbles. If hard, large particles are present at sufficiently high velocities, damage will occur. In contrast to liquid droplet impingement, the necessary velocities for solid particle erosion are quite low. Surfaces damaged by solid particle erosion have a very variable morphology. Manifestations of solid particle erosion in service usually include thinning of components, a macroscopic scooping appearance following the gas/particle flow field, surface roughening (ranging from polishing to severe roughening, depending on particle size and velocity), lack of the directional grooving characteristics of abrasion, and in some but not all cases, the formation of ripple patterns on metals.

strain induced corrosion cracking. Strain induced corrosion cracking is used to refer to those corrosion situations in which the presence of localized dynamic straining is essential for crack formation (i.e. initiation and propagation) to occur, but in which cyclic loading is either absent or restricted to a very low number of infrequent events. Strain induced corrosion cracking has been observed in particular in German NPPs made of higher strength carbon steel and low alloy steel.

stress corrosion cracking. SCC is a localized non-ductile failure which occurs only under the combination of three factors: (1) tensile stress, (2) aggressive environment, and (3) susceptible material. The SCC failure mode can be intergranular SCC or transgranular SCC.

stress oriented hydrogen induced cracking. Arrays of cracks that are aligned nearly perpendicular to the applied stress, which are formed by the link-up of small hydrogen induced cracks in steel. Tensile stress (residual or applied) is required to produce stress oriented hydrogen induced cracking. Stress oriented hydrogen induced cracking is commonly observed in the base metal adjacent to the heat affected zone of a weld, oriented in the through-thickness direction. Stress oriented hydrogen induced cracking may also be produced in susceptible steels at other high stress points such as from the tip of mechanical cracks and defects, or from the interaction between hydrogen induced cracking on different planes in the steel.

---

[2] Adapted from WERMAC, Definition and Details of Socket Weld Fittings ASME B16.11, with permission.

structural reliability model. Sometimes the terms 'PFM' and 'structural reliability model' are used synonymously. While there are some common elements in the underlying methodologies, a structural reliability model is concerned with 'stress-strength' in which it is assumed that a component fails if the applied stress exceeds its strength; called a type II structural reliability model in some classification schemes.

thermal ageing embrittlement. A time and temperature dependent change in the microstructure of the material which leads to a reduced ductility and deterioration of the fracture toughness and the impact properties. The material will show an increased embrittlement over time.

thermal stratification. Hot water can flow above cold water in horizontal runs of piping when the flow (hot water into a cold pipe or cold water into a hot pipe) does not have enough velocity to flush the fluid in the pipe. The temperature profiles in the pipe where the top of the pipe is hotter than the bottom causes the pipe to bow along with the normal expansion at the average temperature.

thermal striping. Incomplete mixing of high temperature and low temperature fluids near the surface of structures with subsequent fluid temperature fluctuations give rise to thermal fatigue damage to wall structures.

through-wall crack instability. A condition where a through-wall crack grows around the pipe circumference in a rapid manner.

Tinel coupling. A pipe coupling made of Tinel material (half titanium and half nickel). The coupling is machined to predetermined dimensions with the inside diameter smaller than the tube's outside diameter. Passing a mandrel through the coupling's inside diameter while submerged in liquid nitrogen at cryogenic temperature will expand the coupling. It will remain expanded while submerged in liquid nitrogen. When the coupling is installed to couple two pieces of tubing together, it will automatically return to its predetermined dimensions when exposed to room temperature. The shape memory is the result of a change in the crystal structure of the alloy known as reversible austenite to martensite phase transformation.

transgranular stress corrosion cracking. Transgranular SCC is caused by aggressive chemical species especially if coupled with oxygen and combined with high stresses.

Vegas algorithm. A method for reducing error in Monte Carlo simulations by using a known or approximate probability distribution function to concentrate the search in those areas of the integrand that make the greatest contribution to the final integral. The Vegas algorithm is based on importance sampling.

visual examination. The oldest and most commonly used non-destructive examination method is visual testing (VT), which may be defined as "an examination of an object using the naked eye, alone or in conjunction with various magnifying devices, without changing, altering, or destroying the object being examined." [43] Per ASME XI, there are three different VT methods; VT-1, VT-2 and VT-3.

von Mises stress. Von Mises stress is a value used to determine if a given material will yield or fracture. It is mostly used for ductile materials, such as metals. The von Mises yield criterion states that if the von Mises stress of a material under load is equal or greater than the yield limit of the same material under simple tension then the material will yield.

VT-1 examination. As specified by ASME Section XI, a limited visual examination which is the observation of exposed surfaces of a part, component or weld to determine its physical condition including such irregularities as cracks, wear, erosion, corrosion or physical damage.

VT-2 examination. Per ASME XI, VT-2 is a visual surface examination to locate evidence of leakage from pressure-retaining components.

VT-3 examination. A limited visual examination specific to ASME Section XI, which is the observation to determine the general mechanical and structural condition of components and their supports, such as the verification of clearances, settings, physical displacements, loose or missing parts, debris, corrosion, wear, erosion, or the loss of integrity at bolted or welded connections. The VT-3 examinations include those for conditions that could affect operability of functional adequacy of snubbers, and constant load and spring type supports. It is intended to identify individual components with significant levels of existing degradation. As it is not intended to detect the early stages of component cracking or other incipient degradation effects, it is not to be used when failure of an individual component could threaten either plant safety or operational stability. This examination may be appropriate for inspecting highly redundant components (such as baffle-edge bolts), where a single failure does not compromise the function or integrity of the critical assembly.

water hammer. If the velocity of water or other liquid flowing in a pipe is suddenly reduced, a pressure wave results, which travels up and down the pipe system at the speed of sound in the liquid. Water hammer occurs in systems that are subject to rapid changes in fluid flow rate, including systems with rapidly actuated valves, fast-starting pumps and check valves.

weld inlay. A mitigation technique defined as application of primary water SCC resistant material (Alloy 52/152) to the inside diameter of a dissimilar metal weld that isolates the primary water SCC susceptible material (Alloy 82/182) from the primary reactor coolant.

welding procedure specification. A formal written document describing welding procedures, which provides direction to the welder or welding operators for making quality production welds as per the code requirements. The purpose of the document is to guide welders to the accepted procedures so that repeatable and trusted welding techniques are used. A welding procedure specification is developed for each material type and for each welding type used.

ABBREVIATIONS

AECL	Atomic Energy of Canada Limited
ASME	American Society of Mechanical Engineers
BWR	boiling water reactor
CDF	core damage frequency
CFP	conditional failure probability
CODAP	Component Operational Experience, Degradation and Ageing Programme
CRP	coordinated research project
DDM	data driven model
DiD	defence in depth
EBS	equivalent break size
EPRI	Electric Power Institute
EXJ	expansion joint
GRS	Global Research for Safety (formerly the Gesellschaft für Anlagen und Reaktorsicherheit mbH)
IAEA	International Atomic Energy Agency
I-PPoF	integrated probabilistic physics-of-failure
ISI	in-service inspection
LOCA	loss of coolant accident
LWR	light water reactor
MWP	maintenance work process
NEA	Nuclear Energy Agency (OECD)
NRC	Nuclear Regulatory Commission (USA)
NURBIM	Nuclear Risk-Based Inspection Methodology (for Passive Components)
OECD	Organisation for Economic Co-operation and Development
OPEX	operating experience
PFM	probabilistic fracture mechanics
PORV	power operated relief valve
PRAISE	Piping Reliability Analysis Including Seismic Events
PROST	Probabilistic Structural Mechanics (PFM code)
PSA	probabilistic Safety Assessment
PWR	pressurized water reactor
PWSCC	primary water stress corrosion cracking
PZR	pressurizer
RCS	reactor coolant system
RIM	reliability and integrity management
ROY	reactor operating year
SCC	stress corrosion cracking
SME	subject matter expert
SSCs	structures, systems and components
WCR	water cooled reactor
WWER	water cooled, water moderated power reactor

CONTRIBUTORS TO DRAFTING AND REVIEW

Abdul, J.A.	Malaysian Nuclear Energy Agency, Malaysia
Ahn, D.-H.	Korea Atomic Energy Research Institute, Republic of Korea
Alzbutas, R.	Lithuanian Energy Institute, Lithuania
Blair, C.	Canadian Nuclear Safety Commission, Canada
Cho, W.	International Atomic Energy Agency
Duan, X.	CANDU Energy Inc., Canada
Heckmann, K.	Global Research for Safety, Germany
Jevremovic, T.	International Atomic Energy Agency
Kee, E.	University of Illinois Urbana-Champaign, USA
Lydell, B.	Sigma-Phase Inc., USA
Mohaghegh, Z.	University of Illinois Urbana-Champaign, USA
Reihani, S.	University of Illinois Urbana-Champaign, USA
Riznic, J.	Canadian Nuclear Safety Commission, Canada
Sakurahara, T.	University of Illinois Urbana-Champaign, USA
Wang, M.	CANDU Energy Inc., Canada
Zammali, C.	Tunisian Electricity and Gas Company, Tunisia

Technical Meetings

Vienna, Austria: 12–14 June 2018, 18–21 June 2019,
9–10 June 2020,16–18 June 2021

Structure of the IAEA Nuclear Energy Series*

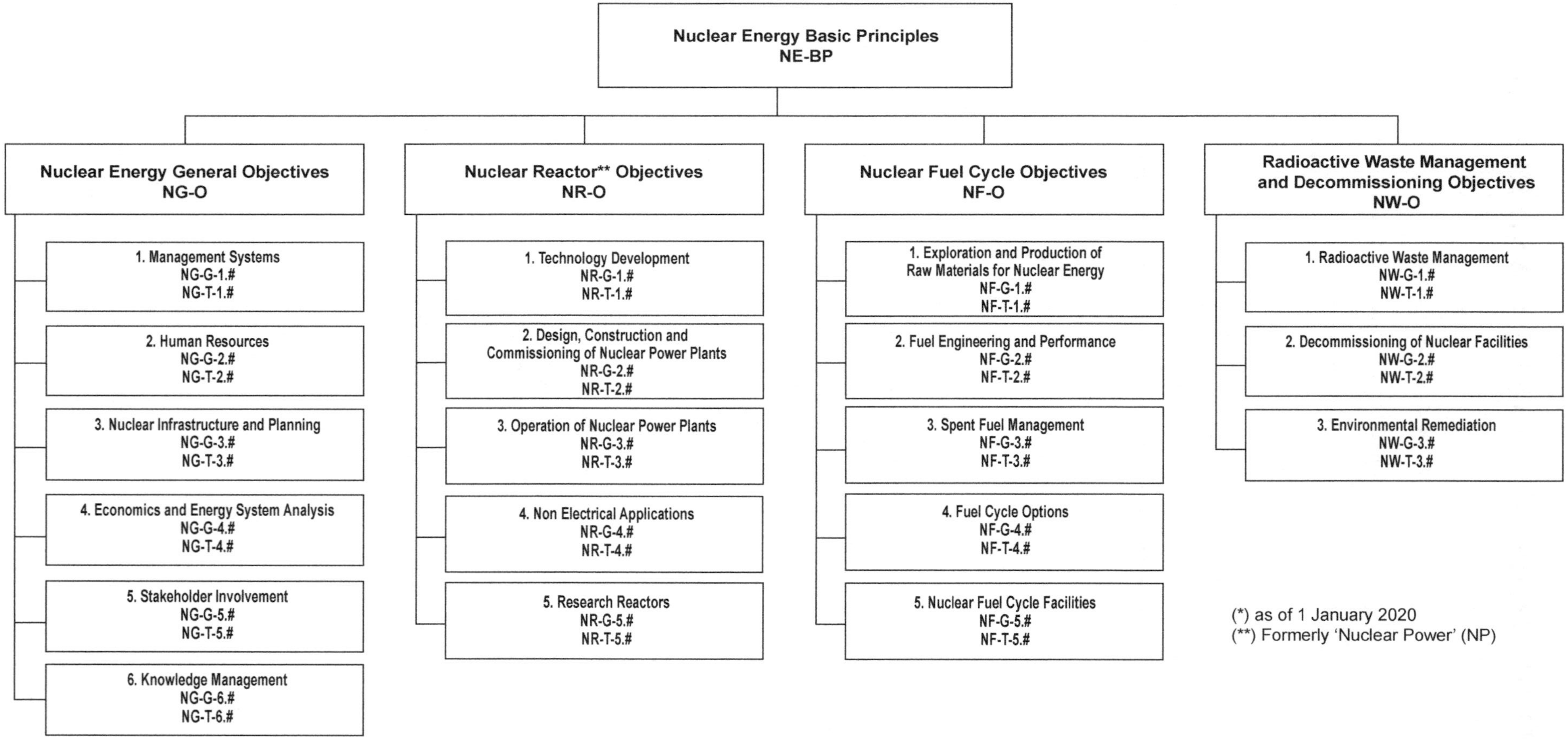

Nuclear Energy Basic Principles
NE-BP

Nuclear Energy General Objectives
NG-O

1. Management Systems
NG-G-1.#
NG-T-1.#

2. Human Resources
NG-G-2.#
NG-T-2.#

3. Nuclear Infrastructure and Planning
NG-G-3.#
NG-T-3.#

4. Economics and Energy System Analysis
NG-G-4.#
NG-T-4.#

5. Stakeholder Involvement
NG-G-5.#
NG-T-5.#

6. Knowledge Management
NG-G-6.#
NG-T-6.#

Nuclear Reactor Objectives**
NR-O

1. Technology Development
NR-G-1.#
NR-T-1.#

2. Design, Construction and Commissioning of Nuclear Power Plants
NR-G-2.#
NR-T-2.#

3. Operation of Nuclear Power Plants
NR-G-3.#
NR-T-3.#

4. Non Electrical Applications
NR-G-4.#
NR-T-4.#

5. Research Reactors
NR-G-5.#
NR-T-5.#

Nuclear Fuel Cycle Objectives
NF-O

1. Exploration and Production of Raw Materials for Nuclear Energy
NF-G-1.#
NF-T-1.#

2. Fuel Engineering and Performance
NF-G-2.#
NF-T-2.#

3. Spent Fuel Management
NF-G-3.#
NF-T-3.#

4. Fuel Cycle Options
NF-G-4.#
NF-T-4.#

5. Nuclear Fuel Cycle Facilities
NF-G-5.#
NF-T-5.#

Radioactive Waste Management and Decommissioning Objectives
NW-O

1. Radioactive Waste Management
NW-G-1.#
NW-T-1.#

2. Decommissioning of Nuclear Facilities
NW-G-2.#
NW-T-2.#

3. Environmental Remediation
NW-G-3.#
NW-T-3.#

(*) as of 1 January 2020
(**) Formerly 'Nuclear Power' (NP)

Key

BP: Basic Principles
O: Objectives
G: Guides and Methodologies
T: Technical Reports
Nos 1–6: Topic designations
#: Guide or Report number

Examples

NG-G-3.1: Nuclear Energy General (**NG**), Guides and Methodologies (**G**), Nuclear Infrastructure and Planning (topic **3**), **#1**
NR-T-5.4: Nuclear Reactors (**NR**), Technical Report (**T**), Research Reactors (topic **5**), **#4**
NF-T-3.6: Nuclear Fuel (**NF**), Technical Report (**T**), Spent Fuel Management (topic **3**), **#6**
NW-G-1.1: Radioactive Waste Management and Decommissioning (**NW**), Guides and Methodologies (**G**), Radioactive Waste Management (topic **1**) **#1**

No. 26

ORDERING LOCALLY

IAEA priced publications may be purchased from the sources listed below or from major local booksellers.

Orders for unpriced publications should be made directly to the IAEA. The contact details are given at the end of this list.

NORTH AMERICA

Bernan / Rowman & Littlefield

15250 NBN Way, Blue Ridge Summit, PA 17214, USA
Telephone: +1 800 462 6420 • Fax: +1 800 338 4550
Email: orders@rowman.com • Web site: www.rowman.com/bernan

REST OF WORLD

Please contact your preferred local supplier, or our lead distributor:

Eurospan Group

Gray's Inn House
127 Clerkenwell Road
London EC1R 5DB
United Kingdom

Trade orders and enquiries:

Telephone: +44 (0)176 760 4972 • Fax: +44 (0)176 760 1640
Email: eurospan@turpin-distribution.com

Individual orders:

www.eurospanbookstore.com/iaea

For further information:

Telephone: +44 (0)207 240 0856 • Fax: +44 (0)207 379 0609
Email: info@eurospangroup.com • Web site: www.eurospangroup.com

Orders for both priced and unpriced publications may be addressed directly to:

Marketing and Sales Unit
International Atomic Energy Agency
Vienna International Centre, PO Box 100, 1400 Vienna, Austria
Telephone: +43 1 2600 22529 or 22530 • Fax: +43 1 26007 22529
Email: sales.publications@iaea.org • Web site: www.iaea.org/publications

22-05658E-T